Urbanisation, Environmental Degradation and Quality of Life

Urbanisation, Environmental Degradation and Quality of Life

H.M. Saxena
M.Z.A. Khan

RAWAT PUBLICATIONS
Jaipur • New Delhi • Bangalore • Guwahati • Kolkata

ISBN 978-81-316-0808-1

Published by
Prem Rawat for **Rawat Publications**
Satyam Apts, Sector 3, Jawahar Nagar, Jaipur 302 004 (India)
Phone: 0141 265 1748 / 7006 Fax: 0141 265 1748
E-mail: info@rawatbooks.com
Website: www.rawatbooks.com

New Delhi Office
4858/24, Ansari Road, Daryaganj, New Delhi 110 002
Phone: 011 2326 3290

Also at *Bangalore, Guwahati* and *Kolkata*

Typeset by Rawat Computers, Jaipur
Printed at Nice Printing Press, New Delhi

Contents

Preface

The present study is the outcome of the U.G.C. major research project entitled, 'Urbanisation, Environmental Degradation and Quality of Life' (A Case Study of One Lakh and Above Cities of Rajasthan). The project has been approved and financially supported by University Grants Commission, Bahadur Shah Zafar Marg, New Delhi.

Several geographical studies have been done on urbanisation and degradation of environment, but their relationship with quality of life has not been explained. The present study is an attempt to investigate the dynamic relationship between urbanisation, environmental degradation and quality of life in cities of Rajasthan having above one lakh population. The subject selected for present study is interdisciplinary in nature and important both from theoretical as well as from applied point of view.

The present study deals with various facets of degradation of environment and urban quality of life. The entire study is based on intensive fieldwork and has been divided into nine chapters. Chapter–I is an introduction which deals with objectives, methodology and review of selected works, while in Chapter–II regional geo-economic and demographic individuality has been explained. Chapter–III deals with trends of urbanisation in Rajasthan. The degradation of environment and spatial pattern of pollution in urban areas have been discussed in Chapter–IV and V. In next two chapters (VI and VII), spatial pattern of quality of life and levels of quality of life have been explained in detail. Chapter–VIII is related with problems and planning, while in last chapter, summary and conclusions have been given. Throughout the study, tables, maps and diagrams have been given to explain the various aspects.

Kirit Jadon has worked as Project Fellow in this project for the period of two years. She has done the entire fieldwork and tabulation work. We are also thankful to officials of the various departments, NGOs and friends who have helped us during the project period in one way or the other.

H.M. Saxena
M.Z.A. Khan

1 Introduction

Nature of Study

At the beginning of the present century, environmental degradation has emerged as a major global concern for human survival. A large number of international and global organisations with the collective wisdom of scientists, economists and planners have come up to settle the environmental issues confronting the nations and the physical world. The environmental crisis has convinced the world to use technology and resources to repair the damage already done to our environment and also to patent substitutes for certain harmful chemicals in order to protect and preserve nature and natural resources. The *Dictionary of Environment* is renewed every time with new terminologies like clean technologies, environmental auditing, environmental friendly products (EFP), etc. Similarly, the concept of environmental management has also emerged and accepted as a tool for sustainable development.

The strains of the ecological crisis are so apparent that the task to preserve and protect the environment has become the primary requisite of the economic development. Modernisation is steadily transforming the nature into products. The resources squeeze has led to an ever intensification of ecological damage. The degradation of environment has now become one of the most challenging problems of the developing countries. It has become more serious in case of urban areas where concentration of population, industries, transportation and other economic activities has given rise to several ecological problems resulting in the lower quality of life.

Urbanisation is, although, a worldwide phenomenon and its impact on environment is more apparent in developing countries like India. In India, including the Rajasthan State, cities are growing at a fast rate and this process will continue to increase because the benefits from high density crowding resulted into, so called 'agglomeration economies' in modern location theory. While having a positive impact on income levels, employment, production and market economics, urbanisation has brought with it many problems, including shortage of housing, inadequate water supply, sanitation and waste disposal facilities, congestion, traffic problems, air, water and noise pollution as well as unsafe social environment. The overall quality of urban environment has deteriorated over the years and resulted in lower quality of life in many parts of the cities.

In India, urbanisation has resulted in serious shortfalls in housing, public utilities and community facilities. The livability in urban places has deteriorated, giving rise to shanty towns and slums, heavy population concentrations, uncontrolled land use, inadequate urban services such as water supply, sewerage and transport, and deficient educational, recreational, health and welfare services. Rapid population growth in urban places has defeated urban goals and planning efforts. In cities where, in some cases, over 30 per cent population lives on pavements and where sub-human conditions prevail there is no effort made uptil now to know about the quality of life. Therefore, there is an urgent need to take into consideration not only the nature of urbanisation but its impact on degradation of environment and the state of quality of life. The present study is designed to discuss the pattern of urbanisation in Rajasthan State and its impact on quality of life in cities having population of one lakh and above.

Objectives

The present study is an attempt to investigate the dynamic relationship between urbanisation, environmental degradation and quality of life in cities of Rajasthan State having population of one lakh and above. The main objectives of the study are as follows:

(i) to unfold the urban environment of one lakh and above population cities of Rajasthan in terms of its physical, economic and demographic structure;
(ii) to explain the nature of urban environment degradation;
(iii) to study the spatial pattern of environmental pollution in terms of air, water, noise and land pollution;
(iv) to analyse the impact of environmental degradation on Quality of Life (QOL);
(v) to explain the spatial pattern of QOL as well as to determine the hierarchy of areas within city based on QOL;
(vi) to trace the problems associated with urban QOL; and
(vii) to suggest a framework for urban planning to control environmental degradation and for the improvement of QOL in cities of Rajasthan.

Hypothesis

In the present study following hypothesis have been tested:

(i) Urbanisation is the main cause of degradation of environment.
(ii) Degradation of environment have had direct impact on QOL.
(iii) Unplanned growth of city is responsible for lower QOL.
(iv) Outer the areas of city, better will be the QOL.
(v) In all the cities, slums are having lower QOL.
(vi) Newly developed colonies have better QOL than older part of the city.

The Concept of Quality of Life

Urbanisation and degradation of environment are well-defined terms, but 'Quality of Life' (QOL) is a new concept at least in geography, therefore, prior to the present study it is necessary to explain the concept of quality of life, particularly QOL in urban areas.

Quality of Life (QOL) is an off-quoted phrase used in various contexts with many shades of meaning. What distinguishes man from the other species is the quality of life, the manner and the extent to which it has changed through history. The history of mankind can be written in terms of evolution of the quality of man's existence. The concept of the quality of life is interdisciplinary in nature and widely studied in sociology, anthropology and economics, similarly several international and national agencies have conducted studies related to QOL for policy making and development. It is not easy to define what quality of life exactly is? In fact, quality of life is a relative term, varies from nation to nation, society to society as well as from person to person. Small and Witherick (1995: 197) have defined QOL as: 'A complex notion concerning the general state or condition of a population in a given area. Undoubtedly, it has an important psychological dimension which takes into account such states of mind as satisfaction, happiness, fulfillment and security. It also has an environmental dimension which embraces such criteria as diet, housing, access to services and safety. Other aspects include considerations such as social opportunity, employment prospects, affluence and leisure time'. The definition given by World Health Organisation Quality of Life (WHOQOL) Group (1995) states: 'QOL is an individual's perception of their position in life in the context of the culture and value systems in which they live and in relation to their goals, expectations, values and concerns ... incorporation ... physical health, psychological state, level of independence, social relations, personal beliefs and their relationship to salient features of the environment. ... quality of life refers to a subjective evaluation which is embedded in a cultural, social and environmental context'. Although, the concept of QOL has been discussed in different ways by social scientists but they have not yet reached on common uniform criteria.

Life has so many aspects and is so diverse in character that richness of one tends to impoverish another. There are several aspects of life as well as quality of life. Some of them are related with philosophy while others with social, economic and development. What is important is: (i) the fulfillment of minimum basic needs, and (ii) adoption of such development strategies which will help in minimising spatial variations between high, medium and low levels of QOL. Thus, the primary aspect of the concept is the identification of parameters which determine the QOL. Several scholars have given their views and have adopted certain indicators in their research on QOL, which not only explains the concept but also provides a base for further studies.

Russell in his Autobiography has enumerated four most preferred values. These are: (i) freedom from ill-health, (ii) freedom from poverty, (iii) good relations with the near and dear ones, and (iv) success in the chosen field of work. The basic quality of life seems to be freedom but at the same time material conditions of the quality of good life are the prerequisite of freedom. Under this category fall such things as economic well-being, healthy environment, social harmony, etc.

Various economic theories and development strategies have at various times expressed six different types of indicators of QOL: (i) per capita income, (ii) growth of per capita income, (iii) quality of life index, (iv) basic needs satisfaction, (v) levels of living, and (vi) Maslowian theory, the hierarchy of needs. Man does not live by bread of GNP alone but he also cannot live without it. A certain minimum growth rate or per capita income or entitlement is a necessary condition for quality of life.

In search of alternative strategies, economists tried to create a measure that could indicate better than per capita GNP, i.e. *Physical Quality of Life Index* (PQLI). The purpose of the exercise was to identify those services which could be provided to the weaker sections. The World Bank has also accepted this concept and funded PQLI programmes from the middle of 70s but today PQLI strategy seems to have disappeared from the corpus of economic thinking. Another strategy developed in this field was *Basic Needs Strategy* (BNS), or the *Minimum Needs Strategy* (MNS). The strategy generally includes primary and adult education, primary healthcare, water supply, nutrition, rural road construction, and housing for all. Some scholars believe in the need for increasing ecology and there is no two opinion about the fact that a good/healthy environment is a prerequisite for a good quality of life. Similarly, others are of the opinion that society's cultural values are an expression of the quality of its people. Maslow has divided needs or wants on the following basis: (a) physical needs, (b) social needs, (c) psychological needs, (d) cultural needs, and (e) spiritual needs. The quality of life is really a function of all these, some of them are measurable, others are not.

Geographical Approach and Review of Important Research Works

The study of QOL in geography is basically related with welfare geography, although several other branches such as social, cultural, urban, economic and environmental geography, etc. have had their concern with this issue. The concern for social well-being of man has found an increasing eco in geography. The decade of the 70s witnessed a swing towards human welfare issues such as poverty, hunger, crime, social and political discrimination, access to social services including health, education and so on – questions which are not just related to development but the quality of life. This was the beginning of the research to measure real differentiation of social well-being including QOL.

The concept of the 'Quality of Life' (QOL) is related with welfare geography, according to Henderson and Quandt (1958: 201), 'The objective of Welfare Geography is the evaluation of the social desirability of alternative geographical states'. While Nath (1973: 11) has described, 'Welfare Geography as a part of geography where we study the possible used several variables for the determination of the levels of QOL in various parts of the world. A brief review of their work will help in determination of a general criterion.

Environmental Protection Agency (1973) has applied a procedure known as 'Delphi' to identify group value judgements with respect to what determines the quality of life. The criterion adopted by Dalkey and Rourke is based on the following variables:

1. Demographic process;
2. Health;
3. Housing;
4. Education;
5. Essential living costs;
6. Public participation;
7. Choices in life;
8. Economic security;
9. Landuse planning; and
10. Economic opportunity.

Berlin Centre for Future Research has developed a criterion for the measurement of 'Quality of life' under the direction of Koelle (1974). The purpose of this study was to arrive at an operational definition of the quality of life by generating a list of weighted goals. These are as follows:

	Primary Goals	*Weights*
(a)	Improvement of the material quality of life	185
(b)	Improvement of the physical quality of life	269
(c)	Improvement of the mental quality of life	288
(d)	Improvement of the spiritual quality of life	258

Each primary goal has been divided into four interim goals, on the basis of above methodology some studies have been done in Germany and some European countries. But the main difficulty with this kind of analysis is that the preferences expressed may be unrepresentative.

Organisation for Economic Co-operation and Development (OECD, 1973) has developed social indicators for human well-being and QOL. The indicators selected by them are:

1. Health;
2. Individual development through learning;
3. Employment and quality of working life;
4. Time and leisure;
5. Command over goods and services;
6. Physical environment;
7. Personal safety and administration of justice; and
8. Social opportunity and participation.

Drewnowski's approach is based on state of well-being which determined the QOL. His well-being index is as follows:

1. Somatic Status (Physical development level)
   (a) Nutritional status
   (b) Health status
   (c) Life expectancy
   (d) Physical fitness
2. Educational Status (Mental development)
   (a) Literacy
   (b) Educational attainment
   (c) Congruence of education with manpower requirements
   (d) Employment
3. Social Status (Social integration and participation)
   (a) Integration
   (b) Participation

The above three major components are reflecting the physical, mental and social status of the people. He has also developed a level of living index having nine indicators. Both these indexes are helpful in determining the quality of life.

Criteria of Social Well-being in USA – In United States, social well-being is determined on the basis of following indicators (Smith, 1973):

1. Income, wealth and employment;
2. The living environment;
3. Health;
4. Education;
5. Social order;
6. Social belonging; and
7. Recreation and leisure.

These elements not only determined the social well-being but also the essence of QOL and this can help in identification of QOL at two different spatial scales, i.e. national and within the metropolis. While Bederman (1974) has used five variables to generate a quality of life index for Atlanta. These variables are health factor, public order factor, housing quality factor, socio-economic factor and density factor.

Determination of Urban QOL in Indian Conditions

It becomes clear from above mentioned studies that variables used by geographers for determination of the QOL are varied in nature and related with physical environment, income, health and nutrition, employment, leisure, educational status, security, social status, etc. All these studies have been done in developed countries, where conditions of urbanisation and standard of QOL is much higher in comparison to developing countries like

India. In India, urbanisation has taken place at a fast rate but living conditions in cities are still not proper in terms of QOL and these conditions are entirely different than urban conditions of the developed countries. Therefore, the criterion used for the determination of QOL cannot be used in original form. This is also because of the fact that in our country detailed data about individuals/family are not available.

A framework for the determination and measurement of the quality of life has been designed for Indian cities (excluding metros) in general and cities of Rajasthan in particular. This framework is based on earlier studies but modified according to Indian conditions. The spatial pattern of the urban QOL can be determined on the basis of following variables:

(a) Housing conditions;
(b) Nature of roads;
(c) Drinking water facilities;
(d) Sewerage/drainage system; and
(e) Waste disposal facilities.

The above mentioned variables have been used in present study for the determination of urban QOL.

All the above factors along with employment, income, literacy, health and leisure determined the quality of life. Although, these factors are considered independently but in totality they not only direct but determined the spatial pattern of the urban quality of life.

Source of Data

The present study is based on collection, compilation and analysis of data obtained from primary and secondary sources. The secondary data have been obtained from various government and semi-government departments. The secondary data from the following departments have been obtained:

1. Directorate of Economics and Statistics, Rajasthan, Jaipur.
2. Rajasthan State Pollution Control Board, Jaipur.
3. Census Department, Jaipur.
4. Town Planning Department (Jaipur, Kota, Bikaner).
5. Public Health Engineering Department (PHED).
6. Urban Improvement Trust (Kota, Bikaner, Sriganganagar).
7. Municipal Councils/Committee (Kota, Bikaner, Sriganganagar, Bharatpur, Tonk).
8. Office of the Urban Infrastructure Development Project.
9. National Information Centre.
10. Department of Forest and Environment.
11. District Information Centres of various districts, etc.

The primary data have been collected with the help of a questionnaire.

Methodology

Like any geographical study, the present work has been carried out with the help of three principal sources, viz., (i) maps, statistics and available written literature, (ii) direct observation of the phenomena associated with degradation of environment and urban quality of life, and (iii) interviews with urban dwellers of the sample cities. The entire work has been done with four perspectives:

(i) analysis of the system based on available data;

(ii) to conduct an extensive fieldwork to collect information through households;

(ii) to observe and collect actual data pertaining to environmental degradation and quality of life; and

(iv) to explain the role of geo-economic and societal factors in the distribution.

All the above four perspectives or the strategies have been integrated to explain the problem holistically as well as logically.

Figure 1.1 *Rajasthan: Cities Having Population One Lakh and Above (2001)*

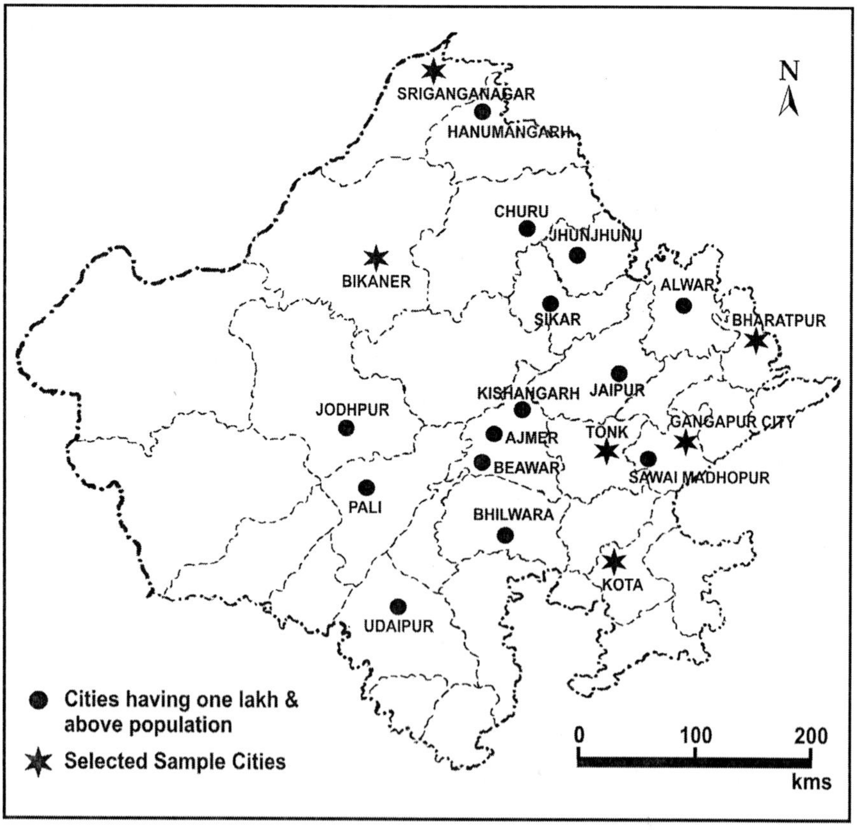

All the collected data have been tabulated and their statistical analysis has been done. Several cartographic techniques have been used to depict the data. The spatial pattern of pollution and QOL in sample towns has been depicted by choropleth method, for this wardwise maps of the selected sample towns have been used. On the basis of tables, maps and diagrams as well as through empirical observations final conclusions have been drawn.

The present study is related with cities of Rajasthan having above one lakh population. The total number of such cities in Rajasthan is 20. It is not possible to study all the cities in limited time, therefore, sampling has been done. In all, five sample cities have been selected, although information regarding other cities has also been obtained, especially for pollution status. The cities selected for sample study are Kota, Bikaner, Sriganganagar, Bharatpur and Tonk. These cities represent different population groups, i.e. Kota and Bikaner having above five lakh population, while Sriganganagar and Bharatpur having population between two to five lakh in 2001. Tonk is in the category of below two lakh population. Gangapur city has also been studied for pollution, it has just above one lakh population. These selected cities also represent different geo-economic regions of the State (Figure 1.1).

The variables selected for the determination of urban QOL are as follows:

(i) Housing conditions,
(ii) Nature of roads,
(iii) Drinking water facilities,
(iv) Internal drainage system, and
(v) Waste disposal facilities.

The spatial pattern of QOL has been determined on the basis of each variable and the general spatial pattern of QOL has been determined on the basis of composite index values and accordingly areas of high, medium and low QOL in each city has been discussed. Its methodology has also been given in concerned Chapter 6.

Coverage

The region selected for the present study is the Rajasthan State (23°4'10" to 30°15'5" North Latitudes and 69°30'5" to 78°16'50" East Longitudes), which is the largest state of the Indian Union with 3,42,239 sq. km area (i.e. nearly 10.43 per cent of India). The Rajasthan State is a unique historic-geographical region having diversified economy primarily based on agriculture, mining, livestock and to some extent on industry. According to 2001 Census 56.47 million population of Rajasthan constituted about 5.50 per cent of India's total population. The percentage of urban population of the State is 23, residing in 222 towns. Detailed geo-economic and demographic characteristics have been described in Chapter 2.

The subject selected for present study is interdisciplinary in nature. Although, our study is based on geographical analysis but it is related with the subjects like sociology, urban planning, economics and environmental studies. Apart from this it will also be helpful to administrators in policy making for the improvement of the quality of life (QOL) in cities.

Another aspect of its interdisciplinary nature is that it will be related with various specialised branches of geography. This is where human geography begins – with the interrelationship of man and environment as well as human well-being. The subject is also related with urban, population, social and cultural geography on the one hand and economic, environmental, physical geography and regional planning on the other. Each of these sub-fields are related directly and/or indirectly with the central problem of the quality of human life. Therefore, this study will be of immense importance both from theoretical as well as from applied point of view.

The entire study has been divided into nine chapters. Chapter 1 is an introduction which deals with nature of the study, objectives, methodology, concept of QOL and review of selected works. While in Chapter 2, geographical individuality of the region has been described. In this chapter physical, economic and demographic characteristics of the state have been described in detail.

Trends of urbanisation in Rajasthan have been described in Chapter 3. It deals with changing pattern of urbanisation in the State of Rajasthan and its present structure.

The problem of environmental degradation has been discussed in Chapter 4. In this chapter firstly general pattern of environmental degradation in Rajasthan has been described and secondly degradation of environment in cities of the State has been discussed. While in Chapter 5 spatial pattern of pollution in urban areas has been described. In this chapter spatial pattern of pollution in Kota, Bharatpur, Sriganganagar, Tonk and Gangapur city has been described in detail.

Chapter 6 deals with the spatial pattern of QOL in urban centres of Rajasthan. In this chapter QOL based on five selected variables has not only been analysed in each sample city but the same has also been depicted on maps. While in Chapter 7 levels of QOL in urban centres of Rajasthan have been determined. On the basis of composite index value area of high, medium and low quality of life in sample cities have been demarcated, in other words hierarchy of the level of QOL has been determined. In Chapter 8 problems related to urbanisation, degradation of environment and quality of life have been discussed and their solutions have been suggested. While in the last chapter summary and conclusions have been described. The entire study is an integrated one and will definitely provide a base for urban planning and improvement of the QOL in cities.

References

Bederman, S.H. (1974), 'The Stratification of "Quality of Life" in Black Community of Atlanta, Georgia', *Southeastern Geographer*, 14, pp. 26–37.

Berry, B.J.L. (1973), *The Human Consequences of Urbanisation*, Macmillan, London.

Coppock, J.T. and C.B. Wilson (1974), *Environmental Quality with Emphasis on Urban Problems*, John Wiley & Sons, New York.

Cox, K.R. (1972), *Man, Location and Behaviour: An Introduction to Human Geography*, John Wiley & Sons, New York.

Dickinson, J.C., R.J. Gray and D.M. Smith (1972), 'The "Quality of Life" in Gainesville, Florida: An Application of Territorial Social Indicators', *Southeastern Geographer*, 12, pp. 121–32.

Drewnowski, J. (1970), *Studies in the Measurement of Levels of Living and Welfare*, UN Research Institute for Social Development, Geneva.

Dutta, Amlan and M.M. Agrawal (eds) (1992), *The Quality of Life*, B.R. Publishing Corporation, Delhi.

Eyles, J. (1974), 'Social Theory and Social Geography', *Progress in Geography*, 6, pp. 27–87.

Fakhruddin (1991), *Quality of Urban Life*, Rawat Publications, Jaipur.

Howe, G.M. (1972), *Man, Environment and Disease in Britain*, David and Charles, Newton Abbot and Barnes and Nobel, New York.

Jana, M.M. (2003), 'Basic Factors for Deteriorating the Quality of Life in India', *Geographical Review of India*, 65 (4), pp. 328–43.

Kayastha, S.L. (1993), 'Environment, Development and Quality of Life', *Annals of the NAGI*, XIII (2), pp. 54–75.

Knox, P.L. (1975), *Social Well-Being: A Spatial Perspective, University Press, Oxford.*

Koelle, H.H. (1974), 'An Experimental Study on the Determination of a Definition for the "Quality of Life"', *Regional Studies*, 8, pp. 1–10.

Liu, B.C. (1973), *The Quality of Life in the United States 1970: Index, Rating and Statistics*, Midwest Research Institute, Kansas City, MO.

Mukherjee, Ramkrishna (1989), *The Quality of Life*, Sage Publications, New Delhi.

Munsi, S.K. (1990), 'Welfare Geography, Poverty and Environment', *Geographical Review of India*, 52 (3), pp. 1–3.

Nath, S.K. (1973), *A Perspective of Welfare Economics*, Macmillan, London.

Nigar, Anjum (1997), 'Habitat, Environmental Degradation and Quality of Life in Modi Nagar', *Geographical Review of India*, 59 (2), pp. 159–63.

Ontell, R. (1973), *The Quality of Life in San Diego*, The Urban Observatory of San Diego, San Diego.

Phillips, David (2006), *Quality of Life*, Routledge, London.

Small, J. and M. Witherick (1995), *A Modern Dictionary of Geography*, Edward Arnold, London.

Smith, D.M. (1973), *The Geography of Social Well-Being in the United States*, McGraw-Hill, New York.

— (1977), *Human Geography: A Welfare Approach*, Edward Arnold, London.

Terleckyj, N.E. (1975), *Improvements in the Quality of Life: Estimates of Possibilities in United States*, National Planning Association, Washington, D.C.

WHO QOL Group (1995), 'The World Health Organisation Quality of Life Assessment: Position Paper from the WHO', *Social Sciences and Medicine*, 41, pp. 1403–09.

Wilson, J.O. (1969), *Quality of Life in the United States: An Excursion into the New Frontiers* of *Socio-Economic Indicators*, Midwest Research Institute, Kansas City, MO.

2 Geographical Individuality of the Region

Regional geographical environment not only directs but also determines the economic, social and cultural development of each region. The various facets of human activities and their interrelated functions, including urbanisation and quality of life are best manifested in the background of the geographical environment which needs to be explained before any study. Therefore in this chapter a brief analysis of physical, economic and demographic conditions has been done. This will help not only in unfolding

Figure 2.1 *Rajasthan: Administrative Units*

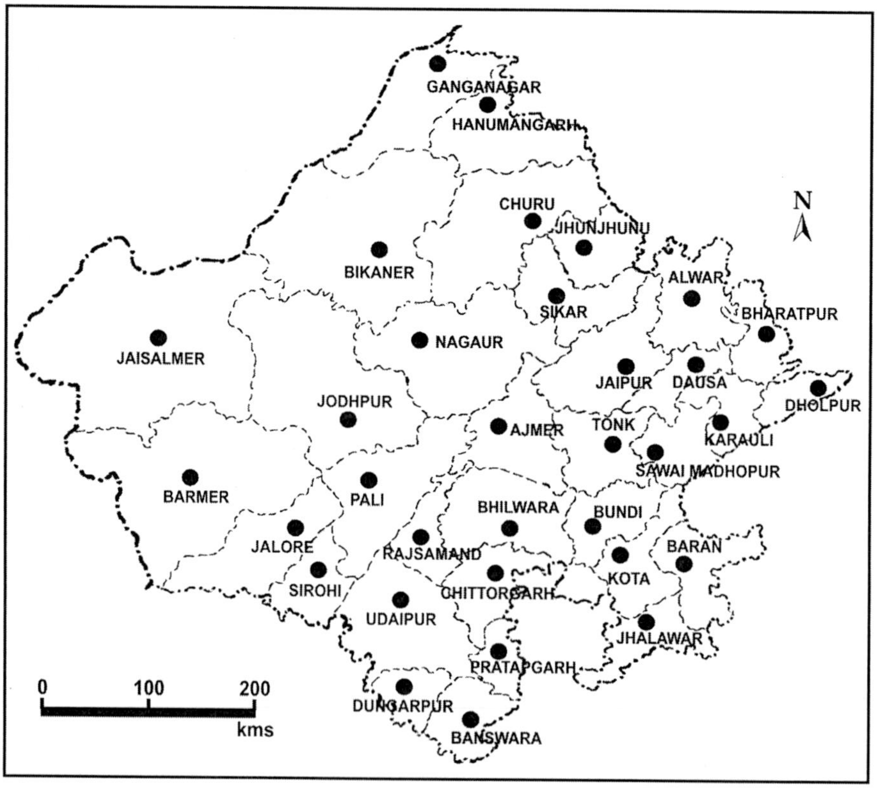

the regional geographical pattern but also in explaining the pattern of urbanisation, degradation of environment and quality of life.

The region, Rajasthan State is a distinct historic-geographical region, located between 23°3' and 30°12' north latitude and 60°29' and 78°17' east longitude. The State covers an area of 3,42,274 sq. km and is situated in the western part of the northern India. Rajasthan is bounded on the west and north-west by Pakistan, on the north and north-east by Punjab, Haryana and Uttar Pradesh, on the south-east by Madhya Pradesh and on the south-west by Gujarat. The State is having 33 districts as administrative units (Figure 2.1).

Physical Environment

Physiography

Rajasthan covers the territory which has a very mature topography, developed during the long years of denudation and erosional processes. The present physiography and drainage system have been greatly influenced and determined by the geological formation and structure. The present landforms are the product of the past fluvial cycle of erosion and recent desert cycle of erosion.

The chief characteristic of the geology and rocks of Rajasthan is noticed in its remarkable succession of Precambrian rocks which go far down towards the base of the Archaean. The sequence commences from the Bundelkhand gneiss, which is one of the oldest granites exposed anywhere upon earth's surface. The early geologist call it a 'fundamental gneiss' of the horizontal and unaltered, but unfossiliferous sediments of the Vindhyan system (Mishra, 1967: 9). In Rajasthan the Aravalli range runs from north-east to south-west. It is considered to be the oldest folded range in the world which still remains in the form of ranges. This mountain chain was peneplaned in Paleozoic times and was subsequently re-uplifted in the Mesozoic Era and this uplifting shows its physical effects in the exceptional local steep folding and faulting of the Vindhyans along the Great Boundary Fault. In brief, the structural pattern of Rajasthan is very complex and there is no need to discuss it in detail. But structurally, the State can be divided into two major divisions. The line demarcating the division runs along the western edge of the Aravalli range which run from north-east to south-west. The south-western part of this line is covered by the north-western extension of the peninsular block. The north-western is also an extension of this block structurally; while surface features indicate rock outcrops at different places, the greater part of the surface is covered by shifting as well as fixed sand deposits.

Physiographically, Rajasthan is a highly diversified State, stretches into two of India's major physiographic divisions, i.e. the Great Plains and the

Central Highlands. The area lying west of the Aravallis known as western Sandy Plain occupies the western part of the Great Plains; while the area east of the Aravallis falls in the northern part of the Central Highlands. Both of these major divisions in Rajasthan are marked by a variety of physiography and relief features (Figure 2.2).

Figure 2.2 *Rajasthan: Physiography*

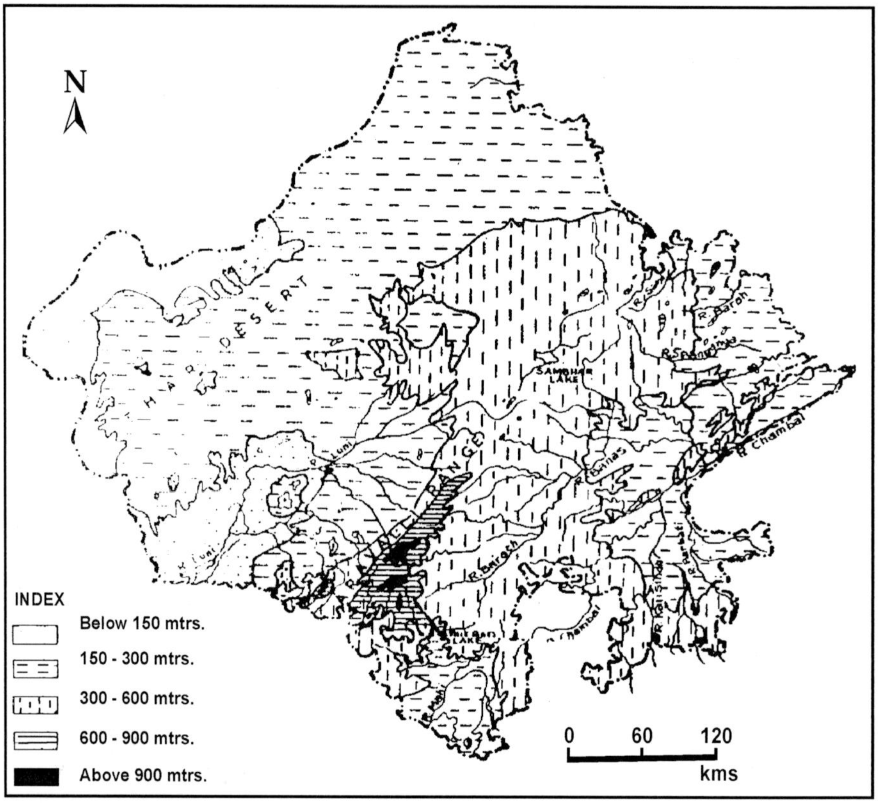

On the basis of the existing relief features, Rajasthan has been divided into the following physiographic divisions:

1. *Western Sandy Plains*
   (a) Sandy Arid Plains (Marusthali)
   (b) Semi-Arid Transitional Plains (Rajasthan Bagar)
      (i) Luni Basin (Godwar Tract)
      (ii) Plain of Interior Drainage (Sekhawati Tract)
      (iii) Ghaggar Plain
2. *Aravalli Range and Hilly Region*
   (a) Aravalli Range and Bhorat Plateau
   (b) North-eastern Hilly Region

3. *Eastern Plains*
   (a) Banas Basin
   (b) Chappan Plains
4. *South-eastern Plateau Region (Hadoti Plateau)*
   (a) Vindhyan Scarpland
   (b) Deccan Lava Plateau

Western Sandy Plains

The western plain covers a large part of Rajasthan located west of Aravalli and is marked by the 50 cm rainfall line. The western sandy plains are further subdivided into two major units: (a) Sandy Arid Plain, and (b) Semi-Arid Transitional Plain.

The Sandy Arid Plain or Marusthali region covers the greater part of the plain of Marwar and it includes the districts of Bikaner, Jaisalmer, Churu, part of western Nagour and western two-thirds of Barmer and Jodhpur districts. This is a part of the Great Indian Desert known as 'Thar Desert'. A large part of this desert is covered by sand dunes, while in some parts of Barmer, Jalore, Jaisalmer rock outcrops are exposed at the surface.

The Semi-Arid Transitional Plain or Rajasthan Bangar runs in north-east to south-west directions. The eastern part of the plain is covered with sand deposits. In this zone the older rocks protrude above the surrounding sandy surface more frequently than in western arid plains. This plain is further subdivided into three smaller physiographic units, i.e. (i) Luni Basin, (ii) Plain of Interior Drainage (Sekhawati Tract), and (iii) Ghaggar Plain.

Aravalli Range and Hilly Region

The Aravalli range, though not of a uniform width, runs diagonally across the State from north-east near Delhi and extends to south-west up to the plains of Gujarat for a distance of about 692 km. Within Rajasthan, the range runs from Khetri in the north-east to Khed Brahma in the south-west for a length of about 550 km. The Aravallis are considered to be the oldest folded mountain range in the world.

From Delhi, the ridges extend to the south-west, but near Khetri they become more prominent and appear in the form of well marked ranges. Towards the south-west, the ranges become more prominent and peaks like Babai (780m), Kho (920m), Ragunathgarh (1,055m) and Taragarh (873m) are located. These ranges pass west of the Sambhar Lake and cover a much wider area. From Ajmer onward several parallel ranges become conspicuous. The Aravalli hills beyond Merwara occupy a width of about 50 km and begin to spread out to the south and south-east towards Udaipur and Dungarpur. Here the elevation is much lower and the principal range extends to the south-west, up to the south-east corner of Sirohi district. Gurushikhar (1,727m), the highest peak in Rajasthan, is located in this hill region in Mt. Abu.

The whole Aravalli range and hilly region has been further subdivided into two smaller physiographic units, i.e. (i) Aravalli range and Bhorat Plateau, and (ii) North-eastern Hilly Track. In Aravalli range region east of Sirohi and Rajsamand, Udaipur and Dungarpur districts are included. The highest section of the Aravalli lies north-west of Udaipur between Kumbalgarh and Gogunda. Locally this section is also known as 'Bhorat' plateau, the average elevation of this plateau is 1,022m. On the other hand north-eastern hilly track lies mostly in Alwar district. In this part, the elevation of these ranges varies between 550m to 670m further north and north-east, the hills are broken and their elevation decreases in the final extension of hills south of Delhi.

Eastern Plains

The eastern plain covers the area north-east, east and south-east of the Aravalli range. This plain is further subdivided into two physiographic units, i.e. (i) The Banas Basin, and (ii) The Chappan Plain.

The Banas Basin covers eastern part of Udaipur, western Chittorgarh, Bhilwara, Tonk, Jaipur, western Sawai Madhopur and southern part of Alwar districts. The plain is drained by the river Banas and its tributaries like Khari, Sodra, Moshi and Morel joining on the left, while Berach, Bajaen and Golwa join on the right.

The Chappan plain lies south of the great Indian watershed in south-eastern Udaipur, Banswara and the southern part of Chittorgarh district. The area is drained by the tributaries of the Mahi river. The Chappan area presents an entirely different set of landforms. The land in the south is deeply and intricately dissected resulting in the formation of separate hills. This deeply dissected area is locally known as '*bangar*' and includes the hilly tracts of Bhilwara and Dungarpur.

South-eastern Plateau or Hadoti Plateau

The south-eastern part of Rajasthan is a plateau region known as 'Hadoti Plateau'. The region covers districts of Kota, Bundi, Jhalawar and Baran. The greater part of this area is drained by river Chambal and its right bank tributaries like Kali Sindh, Parwan and Parbati. Physically this plateau has been divided into two smaller well defined units: (a) Vindhyan Scarpland, and (b) Deccan Lava Plateau. In this region crescent shaped ranges spread from Bundi district to southern part of Kota district. These ranges are known as Bundi hills and Mukandwara hills. In fact, Hadoti Plateau is a north-western extension of the Malwa Plateau.

Climate

The climate of any region can be considered as one of the most important geographical factors which influence all the human activities and its related aspects including quality of life. Climatically, the Rajasthan State can broadly be divided into arid, semi-arid and sub-humid regions. The western Rajasthan consisting of districts Jaisalmer, Barmer, Bikaner, Ganganagar, Churu, Jhunjhunu, Sikar, Nagaur, Jodhpur, Pali and Jalore is under the influence of arid climate. In this arid region rainfall is low (below 50 cms) and highly variable year after year creating inhospitable living conditions for human and livestock population. True desertic conditions prevail in extreme western part of the State. However, there is improvement in rainfall pattern from the west towards the east Rajasthan creating semi-arid conditions in the districts of Jaipur, Ajmer, Tonk, Sawai Madhopur. While Sirohi, Dungarpur, Banswara, Udaipur, Jhalawar, Baran, Bundi districts as well as eastern districts are having humid conditions.

Climatically, the year in Rajasthan has been divided into three major seasons:

(i) The Hot-Weather Season (March to mid June)
(ii) The Rainy Season (mid June to September)
(iii) The Winter Season (October to February)

The Hot-Weather Season (March to mid June)

In Rajasthan, the hot weather season is initiated in the month of March and it prevails from April to June. The temperature rise during this month is somewhat uniform in the whole of Rajasthan. Due to increase in temperature, the atmospheric pressure falls and winds usually are from west to east. As they come after crossing heated Thar desert, they are characteristically dry and warm as well as sandy, locally known as '*loo*' and '*andhi*'. In western part of Rajasthan mainly at Bikaner, Phalodi, Jaisalmer and Barmer, the mean daily temperature goes up to 40°C to 45°C during these months. During summer months dust storms are the most common phenomena in the western arid region and the number of such storms gradually decreases over semi-arid region and regions with higher rainfall.

The Rainy Season (mid June to September)

The temperature, pressure and wind conditions in June indicates that the existing conditions have become more intensified. The northern, north-western and western India continue to get heated till the onset of the monsoon by the middle of June. Although, Rajasthan is in the route of Arabian Sea Branch of Monsoon, but because of the absence of any relief barrier, its not very effective (the Aravalli ranges are parallel to the monsoon winds). On the other hand this part of India is very far from Bay of Bengal Branch of Monsoon. Due to these reasons, Rajasthan receives lesser amount of rainfall. The rainfall during June to September decreases, most markedly from east to west in the State. The 50 cm isohyetal line divides the State into two sections. Mt. Abu in the south receives about 150 cm of rainfall (highest) in Rajasthan. East of Aravallis, nowhere rainfall increases more than 75 cm. Kota has about 70 cm of rainfall in this season. The arid west has a mean rainfall of less than 25 cm.

The Winter Season (October to February)

The cold weather season has been further divided into two subdivisions, i.e. (a) The season of retreating monsoon from October to December, and (b) The cold season from January to February. The temperature in October in the whole of Rajasthan is fairly uniform, the maximum being 33.3°C to 36.1°C and minimum 17.7°C to 21.1°C during December to February, the season is of true winter. The average mean temperature for the month of January ranges from 12°C in northern part of State to 16°C in southern part. Churu often recorded lowest temperature of 1°C or even less. Similar is the case of Mt. Abu. During winters some rainfall has also occurred, which is known as 'Mawat' and very useful for crops like wheat.

Natural Vegetation

Rajasthan has a great variety of natural vegetation ranging from scanty vegetation in the western arid region to mixed deciduous and sub-tropical evergreen forests in the east and south-east of the Aravalli range. The total area under forest in year 2005–06 was 32,627 sq. km which constitutes 9.54 per cent of the total area. Karoli district is having 38.59 per cent forest area, which is highest in the State, followed by Udaipur (36.66%), Baran (32.07%) and Sirohi (31.89%). The district having lowest forest cover is Churu, with only 0.42 per cent, while Nagaur is having 1.36 per cent and Jaisalmer is having 1.51 per cent. Apart from these districts having less than 5 per cent forest area are Barmer, Bikaner, Jodhpur, Ganganagar, Hanumangarh and Jalore.

The forests of Rajasthan have been classified into the following seven types:

(i) Dry Teak Forests
(ii) Anogeissus Pendula Forests
(iii) Mixed Deciduous Forests
(iv) Boswellia Serrata Forests
(v) Butea Monosperma Forests
(vi) Tropical Thorn Forests
(vii) Sub-Tropical Evergreen Forests

The location and distribution of the forest types have been determined by physical factors like relief, climate, soil as well as by biotic factors.

Dry Teak Forests are located mostly in Banswara, Chittorgarh, Rajsamand, Udaipur and Baran districts. *Anogeissus Pendula Forests* are found in an area located mainly in south-eastern part of the Aravalli hills.

Anogeissus Pendula is a tree of dry hot regions. *Mixed Deciduous Forests* are common in Udaipur, Kota, Bundi, Jhalawar, Chittorgarh and Sirohi districts. *Boswellia Serrata Forests* occur in Alwar, parts of Chittorgarh, Udaipur, Sirohi, Ajmer, Jaipur and Jodhpur districts. *Butea Monosperma Forests* are characteristic of black clayey soils and occur mainly in all the valleys where teak occurs. *Tropical Thorn Forests* grow mostly in the lower hill slopes and undulating terrain in Jodhpur, Jaisalmer and Ajmer districts. While *Sub-Tropical Evergreen Forests* are found around Mt. Abu in Sirohi district.

Soils

Soil is one of the most important natural resources and its importance for agriculture cannot be overemphasised. The soils of Rajasthan have been divided into the following seven groups:

(i) Desert Soil
(ii) Grey and Brown (Desert) Soil
(iii) Red and Yellow Soil
(iv) Ferruginous Red Soil
(v) Mixed Red and Black Soil
(vi) Medium Black Soil
(vii) Alluvial Soil

Desert Soil occupies the largest area of Rajasthan. The whole of the State west of Aravalli is covered by the sand, hillocks and rock outcrops. This soil contains a high percentage of soluble salt and has high pH value; it has varying percentage of calcium carbonate and generally is poor in organic matter. The greatest limiting factor is the scarcity of rainfall. Now the irrigation facilities in many parts of this area have been provided through Indira Gandhi Canal Project.

Grey and Brown (Desert) Soil occurs in the districts of Barmer, Jalore, Jodhpur, Sirohi, Pali, Nagaur, Sikar and Jhunjhunu districts.

Red and Yellow Soil covers western part of the districts of Udaipur, Rajsamand, Bhilwara and Ajmer. The soils are poor in carbonate and humus content but have good moisture holding capacity. Three categories of this soil are as follows:

(a) *Sandy Soil* mostly found on rocky outcrops and sand dunes. The soil is sandy to loamy.
(b) *Shallow Soil* also occurs in areas of rock and rocky outcrops. This is a shallow soil of 30 to 120 cm depth lying over rocky strata.
(c) *Dark Medium Heavy Soil* is fairly deep and is sandy loam having yellowish-brown to dark colour.

Ferruginous Red Soil occupies the central and southern part of Udaipur and whole of Dungarpur district. The soil greatly differs in depth and fertility in different areas.

Mixed Red and Black Soil is found in the eastern parts of the districts of Udaipur, Chittorgarh, Dungarpur, Banswara and Bhilwara. The pH of the soil is neutral to alkaline.

Medium Black Soil is commonly found in Kota, Bundi, Jhalawar and Baran districts. Soil in this region is black and deep and some alluviation of finer particles is also noticed.

Alluvial Soil occupies the north-eastern part of the State mainly in Alwar, Bharatpur, Dholpur, Jaipur, Sawai Madhopur, Karauli as well as Ganganagar and Hanumangarh districts.

Economic Structure

Agriculture

Agriculture in Rajasthan is largely dependent on the vagaries of the monsoon. Cereals, millets, oilseeds, pulses, cotton seed, spices, soyabean, vegetables and arid fruits are the main crops grown in the State. The productivity of most of the agricultural commodities is low as compared to the national average. The State, however, ranks number one in the production of bajara, moth, rapeseed and mustard, coriander, guar and cumin. Animal husbandry is also an important source of income for farmers.

Rajasthan is spread over four of the 15 agro-climatic zones of India as delineated by the Planning Commission:

- *Upper Gangetic Plains:* Ganganagar and Hanumangarh districts.
- *Central Plateau and Hill Region:* Ajmer, Alwar, Banswara, Baran, Bharatpur, Bhilwara, Bundi, Chittorgarh, Dungarpur, Dausa, Jaipur, Kota, Pali, Rajsamand, Sawai Madhopur, Karauli, Tonk, Sirohi and Udaipur.
- *Western Plateau and Hill Region:* Jhalawar.
- *Western Dry Region:* Barmer, Bikaner, Churu, Jaisalmer, Jalore, Jhunjhunu, Jodhpur, Nagaur and Sikar.

Since socio-economic factors directly influence agricultural production, the Planning Commission in its report of the committee on *25 Years Perspective Plans for the Development of Rainfed Areas* (1997) divided the country into four agro-economic zones on the basis of certain agro-economic characteristics like level of land productivity, incidence of rural poverty, etc. These agro-economic zones have following features:

Zone–I: areas with relatively high productivity with either high level of irrigation or high assured rainfall and low incidence of poverty.

Zone–II: areas with relatively low productivity, high rainfall, low level of irrigation and high incidence of poverty.

Zone–III: areas with low productivity, low rainfall and high incidence of poverty.

Zone–IV: ecological fragile areas of northern Himalayan belt. North-eastern region and desert areas of Rajasthan and Gujarat.

Rajasthan falls largely in the last three zones, i.e.

Zone–II: Chittorgarh and Jhalawar.

Zone–III: Ajmer, Alwar, Banswara, Bharatpur, Bhilwara, Bundi, Dholpur, Dungarpur, Ganganagar, Jaipur, Jhalawar, Kota, Pali, Sawai Madhopur, Karauli, Sirohi, Tonk and Udaipur.

Zone–IV: The desert districts.

Land Use Pattern

Table 2.1 shows the pattern of land utilisation in Rajasthan.

Table 2.1 *Land Utilisation in Rajasthan*

Category	*Area in Hectares*	*Percentage Share*
1. Forests	26,97,575	7.87
2. Not available for cultivation		
(a) Area under non-agricultural uses	18,34,963	5.36
(b) Barren and unculturable land	24,27,097	7.08
3. Other uncultivated land		
(a) Permanent pasture and other grazing land	17,05,761	4.96
(b) Land under miscellaneous tree crops and groves	20,118	0.06
4. Land excluding fallow land and culturable wasteland	46,11,212	13.46
5. Fallow land		
(a) Fallow land other than current fallow	22,65,039	6.61
(b) Current fallow	19,39,398	5.66
6. Net area sown	1,67,63,765	48.92
7. Total cropped area	2,15,33,809	–
8. Area sown more than once	47,70,044	–

Source: Statistical Abstract Rajasthan–2009, Directorate of Economics and Statistics, Rajasthan, Jaipur, pp. 150–151.

It becomes clear from the above table that 48.92 per cent area of Rajasthan is under net sown area, which shows that agriculture plays a vital role in the State's economy. But forest area in the State is only 7.87 per cent. The percentage of cultural wasteland is 13.46. During last few decades a significant increase in the area under non-agricultural use has been reported. An effective balance between agricultural development and urbanisation needs to be maintained, which will also benefit the environment.

Cropping Pattern

The major crops in Rajasthan are bajara, jowar, maize, moth, kharif pulses, guar, groundnut, cotton, soyabean, and kharif paddy and wheat, gram and rabi mustard. The cropping pattern has changed over the last four decades as a result of development of irrigation potential, production technology, market prices, industrial demand, etc. During last decade there was a negative growth rate in the area under course cereals and kharif pulses, whereas there was a moderate growth in area under groundnut. However, there was an increase in growth rate in area under cotton and a very high growth rate in soyabean among kharif crops. For rabi crops, the compound growth rate was positive for all crops except barley. The cropping pattern is likely to be determined largely by the profitability of each crop, since farmers are now adopting high yielding crops for better income. The crops on which value addition is relatively high will claim a larger share provided the requisite inputs especially irrigation water and remunerative market prices are available.

Agricultural Production

The State has made phenomenal progress in crop production, especially after the green revolution. The area and production of crops like wheat, rapeseed and mustard, rabi pulses, groundnut, soyabean and cotton shown a secular increase from First to the Tenth Plan; while crops like bajara, jowar, sesame, barley and linseed showed a decelerating trend. The following table indicates the area and production of important agricultural products in Rajasthan in year 2006–07.

Table 2.2 *Area and Production of Important Crops in Rajasthan*

Crops	*Area in 000′ hectares*	*Production in 000′ tonnes*
Bajara	4,910	3,440
Maize	1,032	1,118
Wheat	2,565	7,756
Gram	1,011	873
Rape and Mustard	3,096	3,767
Groundnut	306	399
Sugarcane	109	629
Cotton	350	127
Total Food Grains	12,733	14,928
Total Oil Seeds	4,528	5,167
Total Fibres	357	127
Total Condiments and Spices	383	356
Total Fruits and Vegetables	129	485

Source: Statistical Abstract, Rajasthan–2009, pp. 156–168.

Animal Husbandry

Animal husbandry occupies an important place in the economy of Rajasthan. Since more than half area of this State is arid and semi-arid, the livestock forms the main occupation of quite a large number of people. This industry provides subsidiary occupation to the desert inhabitants, particularly the nomadic tribes, who primarily depend upon livestock for their livelihood. Apart from this, in this State, the livestock constitutes the wealth of a cultivator. The total number of livestock according to livestock census 2003 was 491.36 lakh. Among livestock, goats are having highest number 168.08 lakh, followed by cattle (cows) 108.53 lakh, the number of buffalos was 104.13 lakh and sheep was 100.54 lakh. Other livestocks are horses, camels, donkeys, etc. During 1992–97 period Rajasthan recorded a growth of 12.77 per cent in livestock number but in 1997–2007 period a decrease of 10.10 per cent has been recorded.

Minerals

Rajasthan possesses a veritable range of mineral deposits and is an important mining State of India. Among the large variety of minerals it produces or has reserves, the important ones are building stones, barytns, bentonite, calcite, China clay, lignite, gypsum, lead ore, lime stone, felspar, manganese ore, mica, salt, zinc and copper. Although it produces such a vast and diversified range of minerals, the nature of their reserves, distribution in different parts of the States and their marginal deposits, have not made this area as the leading mineral producing area in our country. The State's present importance as a mineral producing State is largely in relation to its resources of non-ferrous minerals like lead-zinc-silver ore, copper ore, strategic minerals like tungsten ore and industrial minerals like mica, steatits and gypsum. Now petroleum and gas has also been discovered in Barmer-Jaisalmer area and its production has started. This will add new importance to the State in national perspective. Rajasthan is the only producer of Jaspar, Garnet, Wollastonite, Emerald in India. It produces 99 per cent of India's zinc concentrates, 93 per cent gypsum, 89 per cent asbestos, 85 per cent soap stone, 77 per cent lead concentrate, 75 per cent rock phosphate, 70 per cent felspar calcite, 50 per cent wulfemite, 36 per cent of copper ore and 22 per cent mica. Apart from this Rajasthan is an important producer of building stones especially marble. At present, there are 19,613 mining areas in the State.

Industrial Development

Until 1960, Rajasthan virtually had no place on the industrial map of India. Some strategies were worked out to increase investment in public sector industries to promote private sector investment in different industrial sectors, leading to the development of industries in the State. However, Rajasthan still ranks quite low among Indian States in terms of industrial development.

In the 1961–71 decade, the linear growth rate in industrial production in Rajasthan was only 2.2 per cent. This increased to 3.9 per cent in 1971–81 and 6.5 per cent in 1981–91. During 1991–2001, industrial production recorded linear growth rate of 10.3 per cent. Thus, industrial production has recorded a steady growth. However, the contribution of the manufacturing sector to National State Domestic Product is still less than 15 per cent.

Development of new industrial investment in any region requires an investor. Friendly environment despite the limited potential of industrial development in the State, the government has taken initiatives to woo investors like concessions in sale tax, land and building tax, octroi, subsidy on new investment and simpler procedures. A new industrial policy 1998 has also been launched in which top priority on infrastructure development has been given. Several departments of the State government are now providing help in industrial development of the State.

During first phase of industrial development Kota emerged as an industrial city with the establishment of numerous large-scale industrial units. Several small-scale industrial units were set up in Bhilwara, Jodhpur, Alwar and Pali while Ganganagar and Chittorgarh grew as agro-processing and cement producing centres respectively.

During 1970–80 industrial development has been occurred in Jaipur, Ajmer, Ganganagar, Jodhpur, Pali, Alwar, Bhilwara and Udaipur districts. Soon four districts Jaipur, Bhilwara, Alwar and Jodhpur emerged as industrial districts with 46 per cent of industrial units. In other districts also industrial growth has been taken place.

In Rajasthan several mineral based industries have been developed. Among them notable ones are Zinc Smelter at Debari (Udaipur), Super Zinc Smelter at Chanderi near Chittorgarh and also at Rampura Agucha in Bhilwara district, Hindustan Copper Unit at Khetri. Cement industry has developed at 19 centres in the State, similarly there are 13 fertiliser producing units in the State. Engineering industry has also developed at Jaipur, Alwar, Bhilwara, Kota, Bharatpur, Ajmer and Pali. Chemical, Sugar and Glass industry have also developed. Some multinational projects such as CEMORE Glass Ltd., MICO Ltd., Copper Automobile Products and Asian Consolidated Ltd. have also been started. The State government is now providing several incentives for the establishment of new industrial units in the State. Similarly due emphasis has been given for the development of the medium small scale units as well as for the development of cottage industries in the State.

Transportation

The economy of a region and its transport infrastructure are closely interrelated. In Rajasthan, a number of important industrial centres have come into existence and this has led to an increase in road length across the State. In 1950–51 the total road length was 17,339 km which has reached 58,350 km in 1990–91. In year 2000–01 the road length in Rajasthan was 87,462 km and in year 2005–06 it was 1,67,132 km.

There are 12 National Highways passing through Rajasthan with a total length of 4,194 km. The State, thus, ranks third in terms of National Highway network. For better connectivity with national highway system in the State, the State government has proposed additional length of new national highway. The total network of State Highway in Rajasthan is 11,263 km. But this network is not evenly distributed across the State. Rajasthan has 20,318 km of district roads and 53,278 km of rural roads. Apart from this 78,079 km roads are various departmental roads. The State government is very particular regarding road development in Rajasthan and has prepared Rajasthan Road Vision–2025 for this purpose. The projects like Mega Highway, Missing Link, Chief Minister Road Plan, etc. are ongoing in the State.

The total length of railway in Rajasthan is 5,837 km (in year 2005–06). In this 3,550.69 km Broad Gauge, 2,200.54 km Metre Gauge and 86.76 Narrow Gauge. The district headquarters of Jhalawar, Banswara, Karauli, Sirohi, Rajsamand and Tonk remain unconnected with any railway line even today.

The regular air transport facility is available at Jaipur, Jodhpur and Udaipur only.

Demographic Structure

Growth of Population

Table 2.3 indicates the pattern of decennial growth of population in Rajasthan.

Table 2.3 *Decennial Growth of Population in Rajasthan*

Year	*Population (Crore)*
1901	1.30
1911	1.10
1921	1.04
1931	1.18
1941	1.39
1951	1.59
1961	2.01
1971	2.58
1981	3.43
1991	4.40
2001	5.64
2011	6.85

The table shows that during a period of 50 years from 1941 to1991, the population of Rajasthan increased from 1.39 crore to 4.40 crore. Thus there was an increase of 3,01,00,000 persons. During the first 40 years, i.e.

from 1901 to 1941, the increase was to the extent of 36 lakh persons. It is interesting to note that the population of Rajasthan increased by 9.8 lakh persons during 60 years of 1901–61 which was more or less equal to an increase by 97 lakh persons in the single decade 1981–91. The population declined in absolute terms during 1911–21 due to epidemics. The decennial growth of population during 1961–71 was 26.4 per cent, which increased sharply to 33 per cent during 1971–81, but declined slightly to 28.4 per cent during 1981–91. In 1991, Rajasthan had a share of 5.2 per cent in the total population of the country. Thus it can be seen that there was a tremendous growth of population in Rajasthan and the reasons behind this may be high birth rate, improved medical facilities and health services, low death rate, proper housing facilities, improved irrigation facilities, introduction of green revolution, low standard of living and low rate of literacy especially in rural areas causing lack of knowledge about family planning. During the earlier decades, the rate was lower because of the unfavourable uncontrolled natural conditions. There was one lower rise during the decade of 1941–51, when the absolute growth was lesser than the preceding one. Succeedingly 1951 onwards, the growth rate was accelerated till 1991. The most phenomenal increase is marked in the decade 1971–81 when the growth rate in Rajasthan was 32.97, which were much higher than in India. The rate of growth in each decade has remained higher to the country average. In the earlier decades, the mortality rates were also higher in the State.

Variations in Growth of Rural-Urban Population

Apart from general growth of population it is also important to study the growth of population in rural and urban areas. This is especially important in context of present study, i.e. urbanisation and quality of life. The following table shows the nature of decadal variation in rural-urban population:

The growth of population in rural areas of the State since 1901 has shown a fluctuating trend. However, it has registered a higher growth rate of 27.49 per cent during the decade 1991–2001 as against 25.46 per cent registered during 1981–91. The State has registered percentage decadal growth rate of 28.41 during the decade 1991–2001 that was more or less similar during the decade 1971–81. The highest percentage decadal growth rate has been recorded during the decade 1971–81, which was 32.97 per cent. During the decade 1991–2001, 11 districts have attained higher growth rates and the remaining 21 have attained lower than that State's growth rate, i.e. 28.41 per cent.

It is evident from Table 2.4 that during the last nine decades, the growth of population in rural areas has been four times whereas in urban areas it has been six times. This happened due to the migrating trend of the rural population towards urban areas. The growth rate in urban areas has shown a fluctuating trend since 1901 just as in rural population. It registered the highest growth rate of 58.69 per cent during the decade 1971–81. Contrary to the growth rates in rural areas it has shown a declining trend from 39.62 per cent registered in 1981–91 to 31.17 per cent during 1991–2001. This decline in the decadal growth rate of urban population is mainly attributed to declassification of seven municipalities in Gram Panchayats.

Table 2.4 *Percentage Decadal Variation of Population from 1901 to 2001 by Rural-Urban Composition – Rajasthan*

Year	*Total*	*Rural*	*Urban*
1901	–	–	–
1911	+ 6.70	+ 8.74	– 4.83
1921	– 6.29	– 7.26	– 0.03
1931	+ 14.14	+ 13.63	+ 17.21
1941	+ 18.01	+ 17.25	+ 22.43
1951	+ 15.20	+ 10.80	+ 39.59
1961	+ 26.20	+ 29.65	+ 11.04
1971	+ 27.83	+ 25.77	+ 38.47
1981	+ 32.97	+ 27.47	+ 58.69
1991	+ 28.44	+ 25.46	+ 39.62
2001	+ 28.33	+ 27.49	+ 31.17

Distribution and Density of Population

The population distribution is an important indicator of the level of economic and social development of a region. Rajasthan has a population of 5,64,37,122 (according to the 2001 census) contributing more than 5 per cent of the total population of the country. It is the largest State in the country but ranks ninth in regard to population with an average density of only 165 persons per sq. km, which was 129 persons per sq. km according to 1991 census. Population distribution in the State is quite uneven. As we go from east to west, the rainfall decreases and so also the population density. The general pattern of population distribution reveals that in the arid region from Bangar to Marusthali, population is highly scattered and clustered around water points while in the semi-arid Bangar tract where intensity of sand dunes is moderate and some irrigation facilities by tanks and wells are available the general density is more than 79 per cent per sq. km. Here also, the north-eastern portion having Sikar and Jhunjhunu has 296–323 persons per sq. km owing to greater facility of irrigation by wells and tube wells. The canal irrigated areas in Ganganagar district have in general density of 164 persons per sq. km. In contrast eastern Marusthali has a density of 50–100 persons per sq. km while the western Marusthali has below 50 persons owing to progressively increasing intensity of sand dunes combined with extremely low and uncertain rainfall.

The western arid zone extending over an area covering 11 districts, namely Bikaner, Ganganagar, Hanumangarh, Churu, Jodhpur, Barmer,

Jalore, Pali, Naguar, Jaisalmer and Sikar. This region covers an area of 61 per cent of the total area of Rajasthan while population is 31 per cent of the total population of the State. Highest population in this region is of Jodhpur with a population of 28,80,777 spread over an area of 22,850 sq. km. The density of Jodhpur is 126 persons per sq. km which is an average one in respect to the district Sikar where the density is 296 persons per sq. km. Here the rainfall varies from 20cm to 50cm thus affecting the distribution to a greater extent.

As a total of 22,87,229 persons inhabit in an area of 7,732 km. Following Sikar, Nagaur, Pali and Jalore are the preceding ones, while on the other hand population of Jaisalmer is just 5,07,999 over an extensive area of 38,401 with a density of 13 persons per sq. km. This part is known as no man's land due to drought and desert conditions. Density of population decreases from east towards west and north towards south. The density of this region is less than 16 persons per sq. km and shows inhospitable environment and is thus also characterised by low density region. (Figure 2.3)

Figure 2.3 *Rajasthan: Density of Population 2001*

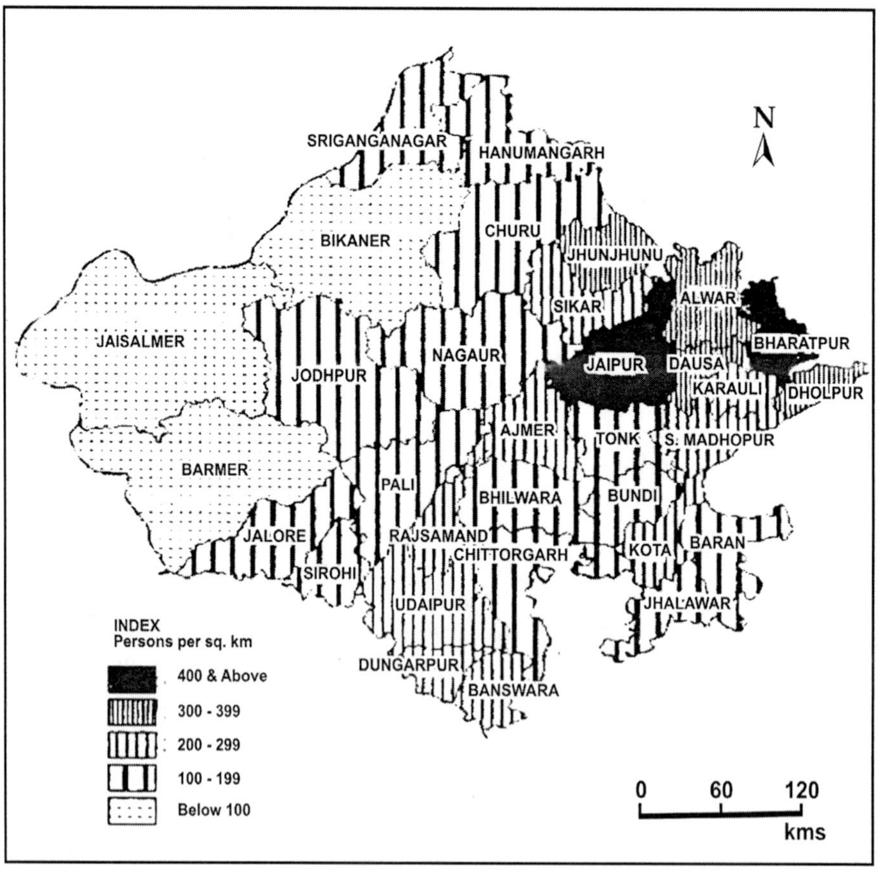

The most striking feature regarding distribution of population in Aravalli region is its concentration in a few favour pockets. The Northern Aravalli with its more or less level topography (except in central part), fertile soil and facilities of irrigation has comparatively more population in the Central Aravalli and is concentrated in and around Jaipur and Ajmer while the Southern Aravalli region with its rugged and hilly terrain and forest cover has spares and widely scattered population. Here the rainfall varies from 50cm to 100cm and sometimes reaches to 150cm. Aravalli comprise districts of Aravalli region which comprise districts of Udaipur, Dungarpur, Banswara, Sirohi and Ajmer covering an area of 9.3 per cent of the total area of Rajasthan, where 10.04 per cent population is inhabited. Udaipur accounts for maximum population of 26,32,210 persons in the region over an area of 12,511 sq. km which is again the largest region among the rest of the regions, while Banswara district accounts for largest density of 298 persons per sq. km followed by Dungarpur with a density of 294 persons per sq. km. In south-eastern region of the State Kota district is having 288 persons per sq. km, while density of population in Chittorgarh district is 166 and Baran 146 persons per sq. km.

Eastern plain region is called the Heartland of the State and includes Jaipur, Tonk, Dausa, Sawai Madhopur, Karauli, Dholpur, Bharatpur, Alwar, Jhunjhunu and Bhilwara. The region possesses maximum of the State with a total of 52,52,388 persons. It covers an area of 23.3 per cent of Rajasthan and has 39.47 per cent of total population. In other words 40 per cent of total population of Rajasthan is found in this region whereas it covers only one-fourth area of the State.

Jaipur is the most populated district not only of this region but also of Rajasthan. If we assume Jaipur as a centre, population around it decreases towards all sides only Bharatpur district stands an exception where density of population is 414 persons per sq. km as against to 471 persons per sq. km of Jaipur. The main reason is that Bharatpur district is situated in Yamuna basin. This region is densely populated because of alluvial level land, good rains, seasonal water supply through rivers and irrigational facilities from dams, etc. Dausa (384), Alwar (357), Dholpur (324) and Jhunjhunu (323) are the other districts having average density of population, whereas Tonk accounts for lowest density of 168 persons per sq. km, while Dholpur has lowest population of just 982 persons per sq. km. Thus we come to know that the concentration, distribution and density of population in the state is greatly affected with amount of rains, nature of land and facilities of irrigation and therefore disparities are seen in regional distribution of population in Rajasthan.

Rural-Urban Population

The rural-urban ratio of population is a significant indicator to measure the level of development of a region. High percentage of rural population shows that the area is still predominantly dependent upon agriculture and that industrialisation has not yet set to engage the population in non-agrarian activities. The urban population is engaged mostly in secondary and tertiary activities and is therefore not directly related to agriculture. The total population of Rajasthan was 564,73,122 in 2001 out of which 4,32,67,678 persons were living in rural areas and the rest in urban areas of the State. Thus the proportion of the population living in rural and urban areas comes to 76.62 per cent and 23.28 per cent respectively.

Table 2.5 *Rural-Urban Composition of Population in Rajasthan since 1901 (in percentage) Proportion of Population*

Year	*Rural*	*Urban*
1901	84.94	15.06
1911	86.56	13.44
1921	85.67	14.33
1931	85.28	14.72
1941	84.73	15.27
1951	81.50	18.50
1961	83.72	16.28
1971	82.37	17.63
1981	78.95	21.05
1991	77.12	22.88
2001	76.62	23.38

Source: Census of India 2001.

The urban component of population of Rajasthan, i.e. the proportion of urban population to its population has been higher than that of India from

Figure 2.4 *Rajasthan Rural-Urban Population 1901–2001*

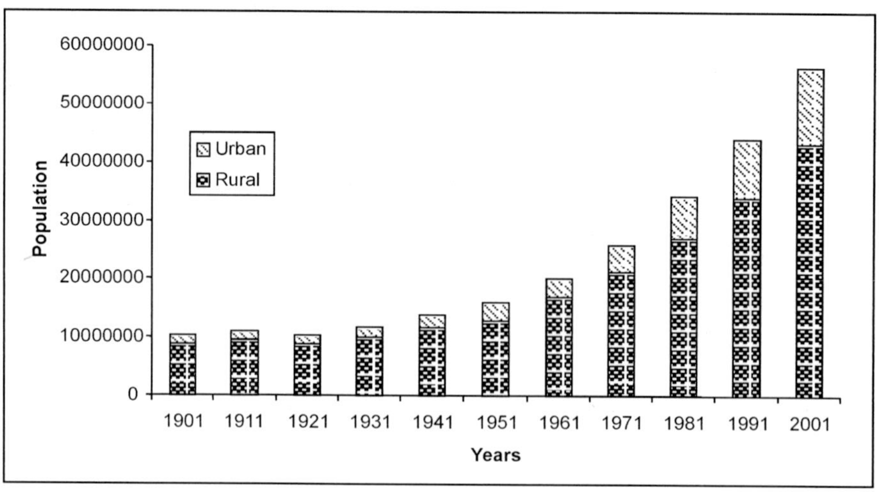

1901 to 1951. After that the proportion of urban population of Rajasthan has been lower than that of India. As regards to the trend, the proportion of urban population after declining from 15.06 per cent in 1901 to 13.44 per cent in 1991, indicated an increasing trend up to 1951 reaching at 18.50 per cent. Again after a decline to 16.28 per cent in 1961 a continuous increasing trend has been registered (Figure 2.4).

In Rajasthan out of 33 districts the percentage of rural population to total population in 25 districts is higher than that of the State average of 76.62. Kota, Jaipur, Ajmer, Bikaner, Jodhpur, Churu and Ganganagar are the districts where the percentage of rural population to total is lower than that of State's average. Kota is the only district in Rajasthan, which can claim to be urban as only 46.58 per cent of population lives in rural areas of the district.

Reference

Mishra, V.C. (1967), *Geography of Rajasthan*, National Book Trust, New Delhi.

3 Trends of Urbanisation in Rajasthan

Rapid population growth and tremendous increase in the number of people preferring to dwell in the cities and towns have been the characteristic feature of the 20th century spatial restructurisation. This growth has been particularly surprising in the developing block of the countries of the world, where urban centres have attracted rural folks in a very large number particularly in bigger cities. Urban centres do provide employment opportunities and many other facilities to their dwellers and to their surrounding zones of influence but the incidence is not homogeneous everywhere. The present trend of urbanisation is more a continuous stratification of people in both the social and economic sense pressurising its potentials and bulging and protruding at many places the cover of their margin. Cities are also termed as the component of economic growth and a creation of the greatest human skill provided they are capable to beget the urbanism in its true sense, and it does not mean it is the only alternative or a compulsion for the poor and the unemployed as it is visible today.

Urbanisation is more a process of transformation both of the geographical landscape and the behaviour of the inhabitants. It is a self-generating process between the interaction of its elements, with their contributory feedbacks and a mutual give and take between its sphere of influence zone. It grows economic growth by spacing economic activities and providing necessary infrastructure. Thus a medium for innovation, diffusion, creation, connection and concentration of the tertiary hub, resulting finally into economic prosperity in the region.

Prior to discussion on trends of urbanisation in Rajasthan, it is necessary to explain the term '*urbanisation*' in geographical perspective.

Urbanisation

Urbanisation not only refers to the changes in the proportion of the population of a nation living in urban areas but also to the process of people moving to cities or other densely settled areas. The term is also used to describe the changes in social organisation that occur as a consequence of population concentration. Urbanisation is thus a process. The process by which rural areas become transformed into urban areas (Palen, 2008: 8–9).

In demographic terms, urbanisation is an increase in population concentration (numbers and density), organisationally, it is an alternation in structure and patterns of organisation. Demographically, urbanisation involves two elements the multiplication of points of concentration and the increase in the size of individual concentrations (Eldridge, 1956: 338).

The process of urbanisation relates to concentration of people engaged in non-agricultural occupation and concentration of non-agricultural land use in a specialised area, a 'place' as a consequence of population, occupational and land-use shifts. Urbanisation thus involves:

1. *Concentration of people* at population densities higher than those associated with agricultural population with only very rare exceptions on either side;
2. *People's shift* (migration) from rural to urban areas; or people shifting;
3. *Occupations shift* from agricultural to non-agricultural; and
4. *Land-use shift* from agricultural to non-agricultural.

The process of urbanisation can be divided into four sub-processes, such as suburbanisation; metropolitanisation – growth of cities with advanced technological and quaternary base; urban-commercialisation – growth of market towns of medium size with an agricultural production and servicing base, with appropriate technology and rural urbanisation – growth of small towns with a service base again with appropriate technology.

Misra (1978: 16) expressed his views on urbanisation as 'a process which reveals itself through temporal, spatial and sectoral changes in the demographic, social, economic, technological and environmental aspects of life in a given society. These changes manifest themselves in the increasing concentration of population in human settlements larger than villagers, in the increasing involvement of the people in secondary and tertiary production and in the progressive adoption of certain social traits which are typical of traditional rural societies'.

Kingsley Davis (1955: 429) points out that 'Urbanisation represents a revolutionary change in the whole pattern of social life. As it is a result of basic economic and technological developments, it tends in turn, once it comes into being, to affect every aspect of existence. Many aspects of the issues given rise to by the rural-urban migration have economic and social implications, so much so that they appear like two sides of the same coin'.

According to the Planning Commission, 'Urbanisation is an important aspect of the process of economic and social development and is closely connected with many other problems such as migration from villages to towns, levels of living in rural and urban areas, relative costs of providing economic and social services in towns of varying size, provision of housing for different sections of the population, provision of facilities like water supply, sanitation, transport and power, pattern of economic development, location and dispersal of industries, civic administration, fiscal policies and the planning of land use'.

Small and Witherick (1995: 255) have explained the main dimensions of urbanisation as: (i) a progressive concentration of people and activities in towns and cities, thereby increasing in general scale of settlement; (ii) a change in the economy of a country or region, whereby non-agricultural activities become dominant; (iii) a change in the structural characteristics of population; (iv) a spread of urbanism beyond the built-up areas of towns and cities, thereby including rural dilution; and (v) the transmission or diffusion of change (economic, social, technology only) down the urban hierarchy and into rural areas.

Ashish Bose (1980: 1) has pointed that the process of urbanisation is a continuing process which is not merely a con-commitment of industrialisation but a concomitant of the whole gamut of factors underlying the process of economic growth and social change. The process of urbanisation is not similar to that of developed and developing countries as Gunnar Myrdal (1968: 471) stated that 'unlike western countries where urbanisation was the symptom of growth caused by economic dynamism of cities, generating the positive demand for labour, urbanisation in South Asia is an aspect of continued poverty. On the one hand urbanisation is a symbol of development but on the other hand it is also a cause of degradation of environment as well as of lower quality of life in certain urban areas'.

Historical Evolution of Urbanisation in the State of Rajasthan

As in other parts of the country, in the Rajasthan State also urbanisation was started during ancient period. The artifacts and Stone Age tools discovered at few sites of Luni Basin, Buddha Pushkar, Ahar Valley, Kalibanga, Piliganga and Sanchore which reflects that the area was inhabited by early man, 6 to 10 thousand years ago. The first urban revolution took place in the area on the banks of river Saraswati and Drishvati in about 2500 BC. This was accompanied by great advancement in human knowledge and technical equipment. These early towns seem to be the product of local increment of concentrated and intensive agricultural activity. They were colonised by the pre-Harappans, later by Harappans and afterwards by several rural communities coming from the north-west. Kalibanga possessed developed culture and the Harappan borrowed town planning fortification, etc. from them.

In the period after the Harappan Age up to the 6th century BC generally no traces of systematic town planning are noticed anywhere and earlier highly developed urban civilisation was succeeded by the rural culture of the tribal people.

In the period from the 5th century BC onwards urbanisation restarted with several small republican and monarchical states such as Malavas, Yaudheyas, Sibis, Arunayans, Maukahris, etc. People gave up nomadic habits and began to settle permanently. Establishment of peace and order resulted in growth of towns. During this period discovery of iron got a revolution in economic sphere. Also, coinage was introduced which stimulated trade and commerce and some towns developed as mint centres.

There were other peculiar factors in growth of towns. Bairath was a Buddhist centre in the 3rd century BC while Pushkar developed as a holy place. Nagari and Nagar were capital towns of the Sibis and the Malavas respectively. Raith flourished as a great metallurgical and commercial centre. Rang Mahal and Hanumangarh grew as route towns for people coming from the north-west. Nails and Rang Mahal became well known for the industry of pottery and clay modeling. Bayana grew as an administrative centre in the 3rd century AD under the Yaudheyas. Jhalarapatan and Gangadhar prospected as they were on banks of the rivers.

The period from the 7th century onwards saw emergence of vast empires of Rajputs who established kingdoms with new capitals. From the 8th century onwards, the appearance of Muslim conquerors caused people to migrate as refugees under the protection of Rajput rulers. The ruling chiefs granted many privileges to these people for habitation of new colonies which they founded. Emergence of new mercantile community which helped in development of trade and commerce brought to the towns. In this Jainism and Saivism made progress and towns became multifunctional in character. Appearance of so many urban centres in desert area is a striking feature in history. These towns must have been like desert ports which served as caravan cities.

Early medieval period saw opening of new lands and new routes and towns sprang up at convenient route places. A large number of capital towns were founded at this time as new kingdoms came into existence and ambitious chiefs carved out separate principalities with new capitals. A large number of Brahminical and Jain holy places came into existence and generally they were situated on spots of natural beauty. Towns gradually sprang up around the religious edifice of the holy places. Prospering commerce and trade under patronage and protection of Rajput rulers allowed small nodal point towns on route ways to flourish and develop as big centres. Pali, Chandrawati, Narhar, Jalore, etc. were prosperous towns of this period. Ajmer, Ranthambore, Nadol and Jalore, Barmer, Hanumangarh, Phalodi, Abu, Bijaulia, Bhillamala, Nagaur and Jaisalmer are some of the notable towns of this period.

The strong traditions of art, architecture and town planning peaked with establishment of Jaipur in 1727 AD wherein the principles of astronomy and ancient Vastu Shastra were utilised for planning of the town.

During the Mughal period the conditions remained almost stagnant. The fall of Mughal empire brought down another setback and broke the economic and social structures of many states patronised under their influence. Trade and commerce between Delhi and Gujarat and Sind survived. The population of Jaisalmer town was 35,000 people and 7,000 household. The number of towns were 24. Jaisalmer contained almost half of the total population of the area, 2 to 3 persons each sq. mile, capital city stands on a peak 254 high, no trees, boundless waste, few stunted shrubs and sea of sands, other centres were Beekampore, Seeruroh, Nachina, Katori, Kahah, Kooldurro, Suthlot, etc. containing on an average 500 to 1,600 people.

The state of Bikaner had 12 urban centres in 1814 having a total of 28,850 houses. Rajgarh was the great commercial mart of this country also a junction of caravan. This area had many ancient towns like Abhore, Bunjaraca, Nuggur, Sadul, Kaliasar, Ratibang, Kalibung, Machotal, etc. The 12 towns along with the number of households were Bikaner (12,000), Nohur (2,500), Bahaderan (2,500), Rinnic (1,500), Rajgarh (3,000), Churu (3,000), Mahjin (800), Jeetpura (1,000), Beedasir (500), Ratangarh (1,000), Sethal (50) and Deshnok (1,000).

In Marwar, river Luni divided this area into fertile and sterile lands. Deedwana, Nagaur, Jodhpur, Pali, Sojat, Godwar, Sewana, Jalore, Bheenmal, Sanchore were the popular towns in this area. Phalodi and Pokaran were important towns in the north. The Desert of (Goga 'Sheo', 'Barmer', 'Kotra' Chohtan) limited the settlements in the south-west.

Every erstwhile state had some political, commercial and trade route centres in the past. Bhilwara in Mewar, Churu in Bikaner, Malpura in Amber, Jhalarapatan in Kota and Pali in Marwar are worth mentioning. State of Amber contains some ancient town sites like Abaner and Mora on the eastern side of Lalsot, ancient capital of Chauhan kings built by the Princes of Dhoondhar, other such sites were Amugarh, Birat, Khandhar near Ranthambore, Ambar, Dotger and Patan. In the north-west the old bed of Ghaggar influenced the old fort towns and their location Anupgarh, Suratgarh and Hanumangarh. Similarly lakes and salt marshes also influenced the growth of towns of which Lunkaransar and Sambhar are examples. The interest, security, segregation as well as cultural and spiritual attachments of Rajput rulers gave birth to cities and towns like Jodhpur, Jaipur, Udaipur, Bharatpur, Sriganganagar, etc. Thus, there were some favourable axis of urbanisation in the past:

(a) Aravalli Hill axis-piedmont and intermontane valleys.
(b) Along major rivers and their tributaries.
(c) Rocky tops and higher slopes.
(d) Fresh water and salt lakes.
(e) Junction of caravan route.

Emergence of the State of Rajasthan

Pre-Independence Scene

Rajasthan faced a perpetual famines of water to feed the people of its then 22 states. The total energy available was only 8 MW and that too unequal everywhere. The economic condition of the peasants was very poor, and the gross productivity had negligible surplus. Rural communities were mostly tradition bound unexposed to outside world against the bleak subsistence agrarian and pastoral background and mass illiteracy. Due to shortage of tertiary occupation and unemployment, urbanisation could not get a much better start. Besides these the physiographic conditions, the expanse of vast unfertile desert soil, scarcity of water for irrigation did not provide a rich background for trade, commerce and early industrialisation.

After the arrival of railways in the last quarter of 19th century a new wave of urbanisation set in. Towns appeared along the railway routes like Phulera, Mantown (Sawai Madhopur), Marwar, Rai-Ka-Bag, Bandikui, Reengus, Mavli, Ratangarh, etc. The volume of trade and commerce also increased with the increase in the number of towns. The growth of urban population during 1901 to 1951 has been given in Table 3.1.

Table 3.1 *Growth of Urban Population in Rajasthan (1901–51)*

Years	*1901*	*1911*	*1921*	*1931*	*1941*	*1951*
Total state urban population	15,50,656	14,75,829	14,75,335	7,29,205	21,17,101	29,55,275
Decadal percentage variation		–4.83	+0.3	+17.21	+22.43	+39.59
Difference over the decadal percentage variation		–	–4.80	+17.18	+5.22	+17.16
Number of urban centres	104	107	115	119	124	147
Percentage of urban population to total State population	15.06	14.43	13.33	14.71	15.27	18.50

Urbanisation During Post-Independence Period

It was only in the post-independence period that the phase of industrialisation, urbanisation and socio-economic transformation got the real start. Expansion in administrative needs, industrial and commercial activities and development in communication gave new impetus to the growth of new towns. Spectacular increase in urban population took place between 1941 and 1951 with a net growth rate of 17.46 per cent from 5.22 per cent between the decades of 1951 to 1961 showing again a tremendous increase of net 20.22 per cent increase in 1981. There was a corresponding increase in the number of towns as well.

After 1961, the State had witnessed fast growth. The urban centres spread into canal command areas. In the semi-arid and tribal belt also small urban centres grew with the expansion of road network. From 1951 to 1991, the State's urban population multiplied by seven times growing from 29.55 lakh to 100.67 lakh.

The urban population of Rajasthan has increased to 13.20 million in 2001 from 10.04 million in 1991, showing an increase of 3.16 million in absolute terms. The percentage decadal growth of urban population during 1991–2001 is 31.17 per cent, as against 39.24 per cent during 1981–91. This implies a slowdown of decadal increase, though not in absolute terms. The percentage of urban population to total population of the State in 2001 stands at 23.88, whereas in 1991 census it was 22.88 per cent. Thus, there has been an increase of 1.0 percentage point in the proportion of urban population in the State during 1991–2001. Table 3.2 indicates the urbanisation level of Rajasthan as compared to other States of India; while Table 3.3 shows urban population growth during 1981–2001 decades.

Table 3.2 *States: Urbanisation Level, 2001*

Highly Urbanised	
Delhi	93.00
More Urbanised	
Goa	49.80
Maharashtra	42.40
Tamil Nadu	43.90
Moderate Urbanised	
Madhya Pradesh	26.67
Rajasthan	23.38
Less Urbanised	
Himachal Pradesh	9.80
Bihar	10.50
All India	27.80

Table 3.3 *Rajasthan: Urban Population and Decennial Growth Rate*

	1981	*1991*	*2001*
Total Population	3,42,61,862	4,38,80,640	5,64,73,122
Urban Population	72,10,508	1,00,40,118	1,32,05,444
Percentage of Urban Population	21.05	22.88	23.88
Share of Urban Population to India's Urban Population	4.52	4.62	4.63
Growth rate of Urban Population	58.69	39.24	31.17
Growth rate of Rural Population	27.47	25.46	27.49

Urban Population Growth

The urban growth of Rajasthan surpassed the country's decadal urban growth rate in last consecutive decades. Invariably, urban growth is faster than rural population growth rate since 1921. The annual growth rate is recorded 2.71 for urban areas.

The State urban population has taken six decades (1960–61) to become 3.28 million from 1.55 million and now two decades to double every time. The fast urban growth is a major concern for both planners and policy makers.

The figures for urban population, the percentage thereof to total population and number of towns are given in Table 3.4 here under:

Table 3.4 *Growth of Urban Population in Rajasthan*

Year	*No. of Towns*	*Urban Population*	*Total Population*	*Percentage of Urban Population*
1901	135	15,50,656	1,02,94,094	15.11
1911	138	14,75,829	1,09,83,509	13.40
1921	147	14,75,335	1,02,92,648	14.30
1931	150	17,29,205	1,17,47,974	14.70
1941	157	21,17,101	1,38,63,859	15.30
1951	227	29,55,275	1,59,70,774	16.20
1961	145	32,81,478	2,01,55,602	16.30
1971	157	45,43,761	2,57,65,806	17.63
1981	201	72,10,508	3,42,61,862	21.04
1991	222	1,00,67,113	4,40,05,990	22.88
2001	222	1,32,05,444	5,65,07,188	23.38

The average annual growth rate in urban-rural have a declining trend after 1981, as shown in Table 3.5.

Table 3.5 *Urban and Rural Annual Exponential Growth Rate*

Year	*Average annual growth rate: Urban*	*Average annual growth rate: Rural*	*U-R growth differential*
1961–71	3.25	2.29	0.96
1971–81	4.62	2.43	2.19
1981–91	3.31	2.24	1.07
1991–2001	2.71	2.01	0.70

Spatially, in the State, there are two types of areas which experienced fast urban growth:

1. The newly carved out districts like Baran, Hanumangarh, Karauli, Rajsamand and Dausa experienced fast urban population growth due to increased administrative activities.
2. The districts in which fast growing city is located like Jaipur, Kota and Bhilwara.

The districts recorded slow growth during 1991–2001 combined two types of contrasting districts:

(a) the districts where rural infrastructure improved therefore commuting is common like Jhunjhunu, Sikar and Churu; and

(b) backward and peripheral districts where limited urban activities developed like Banswara, Jhalawar and Barmer.

Urban Population by Size of Town

The nature of urbanisation becomes clear from the size of towns. The size of town clearly indicates that more the number of million cities, more will be the urbanisation. Table 3.6 indicates the growth pattern of the cities and towns in Rajasthan right from 1901 to 2001.

Table 3.6 *Growth Pattern of Categorywise Cities/Towns in Rajasthan (1901 to 2001)*

Census Year	No. of Towns	Total Population	Number of Towns According to Size and their Percentage					
			I	II	III	IV	V	VI
1901	135	15,50,656	1 (10.33)	4 (15.74)	8 (16.21)	26 (21.88)	64 (27.77)	32 (8.07)
1911	138	14,75,829	1 (9.29)	3 (13.64)	9 (17.68)	24 (21.60)	60 (26.97)	41 (10.82)
1921	147	14,75,335	2 (15.84)	2 (9.69)	7 (14.82)	20 (18.26)	58 (26.77)	58 (14.62)
1931	150	17,29,205	2 (15.25)	2 (10.45)	9 (16.99)	26 (20.83)	64 (26.20)	47 (10.28)
1941	157	21,17,101	4 (27.26)	2 (5.38)	13 (17.69)	28 (18.30)	74 (25.01)	36 (6.36)
1951	227	29,55,275	4 (26.58)	4 (8.92)	20 (19.11)	36 (15.93)	96 (21.74)	67 (7.72)
1961	145	32,81,478	6 (37.84)	4 (7.35)	23 (20.34)	52 (21.56)	5 (11.87)	9 (1.04)
1971	157	45,43,761	7 (40.94)	7 (10.75)	31 (20.03)	67 (20.62)	41 (7.30)	4 (0.36)
1981	201	72,10,508	11 (45.85)	10 (9.18)	55 (23.00)	101 (19.21)	23 (2.70)	1 (0.06)
1991	222	1,00,67,113	14 (49.54)	20 (13.64)	74 (21.72)	87 (13.00)	25 (2.03)	2 (0.07)
2001	222	1,32,05,444	20 (57.23)	26 (13.94)	90 (20.80)	59 (6.92)	17 (1.00)	4 (0.11)

It becomes clear from the Table 3.6 that decadal growth of class–I towns has increased from 40.94 per cent in 1971 to 57.23 per cent in 2001 and their number has also increased from 7 in 1971 to 20 in 2001. Therefore, it is observed that urban process is continuously in favour of bigger cities. This shows the weak impact of all planned intervention of urban programmes related to decentralisation. Out of 20 cities Jaipur is the fastest growing city whereas Ajmer and Beawar are slowest growing cities of Rajasthan. Although, Ajmer city's growth during 1991–2001 in comparison to 1981–91 was high. It was primarily due to provision of assured water supply from Bisalpur drinking water project. The north-eastern and northern cities growth is faster than cities located in western and southern parts of the State during 1991–2001.

In 1951, nearly 45 per cent urban population was living in small towns (< 20,000 population) whereas now it remained only 28 per cent. Census has declared 17 places as census towns. During 1991–2001, the State was very conscious in giving municipal status to any settlement which has population less than 15,000. There are 11 towns below 10,000 population. Except one all are census towns.

There are 20 class–I towns (called cities) having more than one lakh population. Table 3.7 indicates the population growth of towns above one lakh population in Rajasthan.

Table 3.7 *Population Growth of Cities (1,00,000+) in Rajasthan*

	Persons	*1981–1991*	*1991–2001*
Jaipur	2,324,319	+49.56	+59.37
Jodhpur	8,56,034	+31.59	+28.48
Kota	7,04,731	+50.00	+31.14
Bikaner	5,29,007	+44.69	+27.08
Ajmer	4,90,138	+7.22	+21.71
Udaipur	3,89,317	+32.67	+26.17
Bhilwara	2,80,185	+50.02	+52.30
Alwar	2,65,850	+44.14	+26.51
Ganganagar	2,22,833	+30.55	+37.99
Bharatpur	2,05,104	+49.02	+30.74
Pali	1,87,571	+45.44	+37.07
Sikar	1,85,506	+44.00	+25.11
Tonk	1,35,663	+29.08	+35.34
Hanumangarh	1,29,654	+37.73	+56.71
Beawar	1,25,923	+18.58	+17.99
Kishangarh	1,16,156	+32.11	+41.74
Gangapur City	1,05,336	+49.67	+52.91
Sawai Madhopur	1,01,994	+31.49	+31.28
Churu	1,01,853	+33.48	+22.93
Jhunjhunu	1,00,476	+53.01	+39.19

The seven divisional headquarters, namely Jaipur, Jodhpur, Kota, Bikaner, Udaipur, Ajmer and Bharatpur concentrated 40 per cent urban population of Rajasthan. The cities located in eastern and northern parts of the State have grown faster than the cities located in western part. The impact of connectivity of Delhi is clearly visible on the process of Rajasthan urbanisation. Jaipur, Jodhpur and Kota have maintained first three positions in the total hierarchy of towns.

Jaipur has grown much more rapidly in the recent past as compared to the 70s. The Sanganer and Amer townships area merged with Jaipur Municipal Corporation (JMC). The area of JMC increased to 200.40 sq. km in 2001. Several industrial estates have merged with residential colonies of JMC and new are developed in the peripheral like Sitapura, Sudarshanpura, Vindyayka, Kukus, etc. Jaipur has become the heritage city and tourist place of national and international importance. Jaipur is directly linked up by rail with major cities of India, i.e. Delhi, Kolkata, Chennai, Bangalore, Hyderabad, Mumbai, Ahmedabad. Recently, Jaipur airport has been declared as an international airport.

With the passage of time, the small towns have become medium-sized towns and medium became large cities. In 1901, the State was dominated by small towns, whereas at the turn of the 21st century, in the state the small-sized towns are only 4 per cent. At present, the urban process is in favour of large cities. Small towns contain only 28 per cent population. Jaipur city has created its supremacy over all other cities of Rajasthan.

Urban Growth Patterns and Levels of Urbanisation

In Rajasthan the percentage of urban population has also shown an increasing trend. This signifies an increasing gravitation towards urban centres. In the past decade alone about 26 lakh persons were added to the urban centres showing an increasing trend for urbanisation. The urban-rural demographic balance is thus undergoing a process of significant change. This trend is in conformity with the situation in other States of the country. The reason for this is quite obvious. In spite of a number of developmental activities and schemes for employment and occupation generation in rural areas, the fact remains that growth opportunities in rural areas have remained limited. Industrial and social sector development, be it education, health, tourism or any other sector, also remains concentrated in urban areas with better opportunities and facilities in towns and cities. This has been responsible for migration of people from rural areas to the cities in search for jobs and opportunities leading to growth of the urban centres. Another reason has been the growing number of landless and marginal farm lands in agricultural operation in rural areas which with the increase in population in general are finding it difficult to sustain livelihood on practically non-existent land holdings in rural areas and therefore gravitate to the cities and towns in search of livelihood. This situation is going to increase with further addition in population.

Table 3.8 indicates the districtwise levels of urbanisation in 1971, 1991 and 2001, while Figure 3.1 (A) and (B) shows urban population percentage to total population in 1971 and 2001 respectively.

Table 3.8 *Districtwise Levels of Urbanisation and Growth of Urban Population*

S. No.	Districts	Levels of Urbanisation			Growth of Urban Population	
		1971	1991	2001	Combined Annual Growth (1971–2001)	Decadal Growth (1991–2001)
1.	Ajmer	37.65	40.69	40.09	2.38	24.24
2.	Alwar	9.12	13.95	14.53	4.19	35.65
3.	Banswara	5.07	7.72	7.15	3.99	20.32
4.	Baran	–	–	16.84	–	40.73
5.	Barmer	7.26	10.04	7.40	3.22	0.75
6.	Bharatpur	13.76	19.42	19.46	3.56	27.35
7.	Bhilwara	11.03	19.53	20.60	4.33	33.29
8.	Bikaner	41.38	39.73	35.54	3.11	23.56
9.	Bundi	14.60	17.36	18.65	3.42	23.79
10.	Chittorgarh	10.36	15.61	16.04	3.68	24.81
11.	Churu	29.58	28.90	27.87	2.46	20.10
12.	Dausa	–	–	10.31	–	29.05
13.	Dholpur	–	17.19	17.96	–	36.96
14.	Dungarpur	5.89	7.30	7.30	3.22	25.58
15.	Ganganagar	16.48	21.05	25.34	4.05	35.28
16.	Hanumangarh	–	–	20.00	–	39.33
17.	Jaipur	30.05	39.53	49.36	4.42	46.17
18.	Jaisalmer	14.60	15.56	15.03	3.89	44.54
19.	Jalore	4.42	7.28	7.59	4.48	32.18
20.	Jhalawar	9.45	15.78	14.25	3.57	11.43
21.	Jhunjhunu	17.44	20.54	20.65	3.02	21.50
22.	Jodhpur	31.95	35.50	33.85	3.31	27.61
23.	Kaurali	–	–	14.21	–	35.79
24.	Kota	24.05	36.43	53.46	4.43	35.86
25.	Nagaur	12.28	15.98	17.20	3.82	39.24
26.	Pali	11.18	21.75	21.47	4.37	20.84
27.	Rajsamand	–	–	13.04	–	30.96
28.	Sawai Madhopur	11.90	14.84	19.04	3.37	40.17
29.	Sikar	17.03	21.03	20.65	3.32	21.79
30.	Sirohi	17.87	19.51	17.73	2.33	18.15
31.	Tonk	17.45	19.53	20.89	2.84	32.92
32.	Udaipur	12.30	17.10	18.62	3.48	23.85

It becomes clear from the Table 3.8 and Figure 3.1 that districtwise pattern of urbanisation has changed where we take into consideration the urban population in 1971 and 2001. In 1971 Bikaner district was having highest 41.38 per cent of urban population, followed by Ajmer 37.65 per cent, Jodhpur 31.95 per cent and Jaipur district 30.05 per cent. On the other hand Jalore district is having lowest urban population percentage, i.e.

Figure 3.1 *Rajasthan Urban Population*

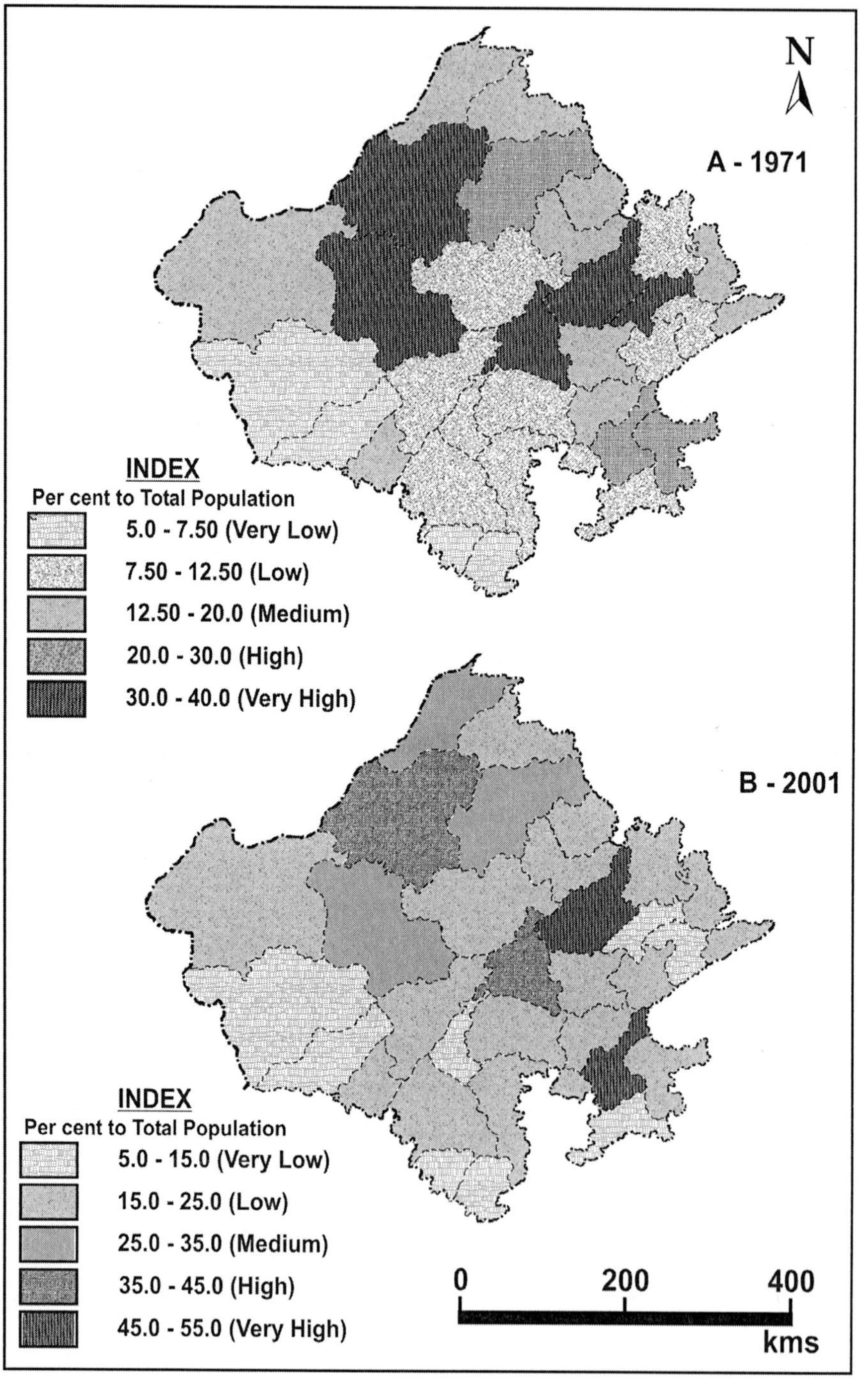

4.42 per cent. Other districts in very low urbanisation category were Banswara (5.07), Dungarpur (5.89) and Barmer (7.26). Jaisalmer, Ganganagar (including Hanumangarh), Bharatpur (including Dholpur), Bundi, Sikar and Jhunjhunu districts are having medium level of urban population between 12.5 to 20.0 per cent urban population. Kota (including Baran) district and Churu was having 24.05 and 29.58 per cent of urban population respectively.

During 1991–2001, five new districts are carved out in the State. Within the State, Kota district is highly urbanised with 53.42 per cent urban population, followed by Jaipur district (49.38%) and Ajmer (40.09%). All the three districts have location of big cities, which are centres of education, health, administration and tourism. Jaipur, Kota and Ajmer are well connected with broad gauge railway lines and also have better road connectivity.

Table 3.9 *Districts: Urbanisation Level*

Urbanisation Level	*Population*
Highly Urbanised	
Kota	53.42
Jaipur	49.38
More Urbanised	
Ajmer	40.09
Bikaner	35.52
Less Urbanised	
Barmer	7.40
Dungarpur	7.24
Banswara	7.15
Rajasthan	23.88

The urban population is lowest in Banswara 7.15 per cent, followed by Dungarpur (7.30%), Barmer (7.40%) and Jalore (7.59%). All the three districts are in the periphery location in terms of Rajasthan State. These districts have high proportion of tribal and backward population, therefore, urbanisation level is below 10 per cent.

The decadal growth rate of urban population has been depicted in Figure 3.2. The highest growth of urban population has been registered in Jaisalmer (44.54%) followed by Jaipur (46.17%) during 1991–2001. While the lowest growth rate was in Barmer district with only 0.75 per cent growth of urban population. The high growth rate of urban population has also been registered in Kota, Baran, Bhilwara, Tonk, Sawai Madhopur, Dholpur, Karauli, Alwar, Nagaur, Ganganagar and Hanumangarh districts in 2001.

Economic Generators Promoting Urban Growth and Process of Urbanisation

In Rajasthan, urban growth is largely based on the following economic generators:

Figure 3.2 *Rajasthan Decadal Growth Rate of Urban Population 1991–2001*

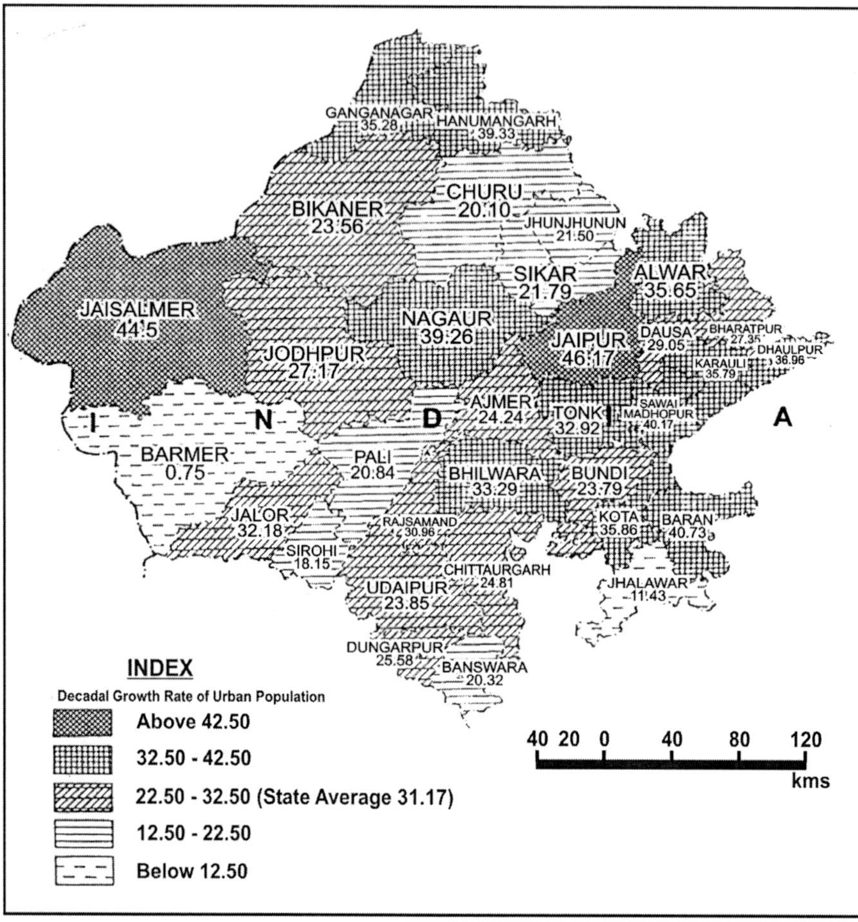

(a) *Agriculture*: The districts in which irrigation potential from the Bhakra Canal System, the Gang Canal System and the IGNP Canal System, has been created are Ganganagar, Hanumangarh, Bikaner, Jaisalmer and Barmer. In the districts of Kota, Bundi, Jhalawar and Baran, the Chambal Command Canal System has benefited these areas. Similarly the Mahi Canal System has benefited Banswara and Dungarpur districts. All these areas have, therefore, become very important growth centres in agricultural and agro-based industries.

(b) *Minerals*: Practically, the entire area of the State, in some way or the other, has the potential for mineral exploration and this has led to activities in mining and industries based on mining in all districts. Particularly important are the districts of Jaisalmer, Udaipur, Bhilwara, Chittorgarh, Rajsamand, Kota, Karauli and Sawai Madhopur.

(c) *Industries*: With increasing power availability and the mineral and agricultural activities, industries have come up in all districts. The major urban centres for industrial activity are Kota, Jaipur, Jodhpur, Ganganagar, Hanumangarh, Udaipur, Bikaner and Pali. One important adverse fall-out of the growth of industries in the State has been the deterioration in environment due to affluent discharge and smoke emissions from industries and power plants. Jaipur, Kota, Bhiwadi, Pali, Udaipur and to a lesser extent practically all other towns having chemical, dyeing, printing, alcohol producing industries are creating pollution and facing problems. Similarly, thermal power plants, such as in Kota, is a great nuisance with ash discharge in a very large area over the city and the problem for ash disposal. There is a need therefore to establish industries with greater caution and with proper safeguards for pollution control and treatment.

(d) *Tourism*: Tourism is an important activity in the State. The total annual tourist arrivals from 1985 to 1989 increased from 33.90 lakh in 1985 to 42.53 lakh in 1989. This has increased during 1996 to 60.70 lakh. The cities having high tourist inflows are Jaipur, Udaipur, Jaisalmer, Jodhpur and Bikaner apart from various medium and small towns. Important tourist circuits have been identified in the State, the prominent being the following:

   (a) Desert Triangle Circuit – Jodhpur, Bikaner and Jaisalmer
   (b) Mewar Circuit – Udaipur, Rajsamand, Pali and Sirohi
   (c) Ajmer, Pushkar Circuit – Ajmer
   (d) Jaipur
   (e) Hadoti Circuit – Kota, Jhalawar, Bundi and Baran
   (f) Sekhawati Circuit – Sikar, Jhunjhunu and Churu
   (g) Mount Abu

These circuits have been drawing large number of tourists with consequent growth of urban centres in the influence areas for providing services.

(e) *Trade and Commerce*: With the increase in economic activities as mentioned above, trade and commerce which had traditional roots amongst people of the State continued to flourish and contribute to the economy. Trade and commerce is the dominating economic activity in most of the urban centres of the State. In development of the trade and commerce road development has played a major role. The trading centres are mainly based on agriculture produce, mining and mineral based products, traditional crafts and industrial produce.

The process of urbanisation is determined by (i) environmental conditions, (ii) agricultural development, (iii) mining development, (iv) industrial development, (v) development of amenities, (vi) infrastructural development, (vii) financial development, (viii) institutional development and (ix)

Table 3.9 *Districtwise Analysis of Urban Process in Rajasthan*

S. No.	District	Environmental	Agricultural	Mining	Industrial	General Facilities	Infrastructural Facilities	Finance	Institutional	Social	Composite Index	Composite Urbanisation	Level of Urbanisation in 2001	Decadal Growth of Urbanisation (1991-2001)
1.	Ajmer	1.091	0.748	0.769	1.087	1.419	1.482	1.385	1.179	1.030	1.132	1.389	40.092	24.240
2.	Alwar	1.085	1.29	0.009	1.965	1.159	1.166	1.278	0.942	0.842	1.082	0.969	14.534	35.650
3.	Banswara	0.996	0.809	0.013	0.674	0.562	0.770	0.575	0.776	0.427	0.623	0.525	7.150	20.320
4.	Baran	1.006	1.667	0.000	2.632	0.839	0.708	0.676	0.902	0.778	1.023	1.112	16.837	40.730
5.	Barmer	0.967	0.858	0.780	0.524	0.490	0.569	0.660	0.909	0.819	0.731	0.192	7.400	0.750
6.	Bharatpur	1.086	1.034	0.097	0.639	0.903	0.900	0.666	0.754	1.068	0.794	0.944	19.464	27.350
7.	Bhilwara	0.944	0.836	3.519	1.178	0.828	1.189	0.910	1.345	0.697	1.272	1.074	20.601	33.290
8.	Bikaner	0.901	1.052	0.386	1.111	1.377	1.262	1.175	1.481	2.209	1.217	1.267	35.540	23.560
9.	Bundi	1.015	1.28	0.467	0.885	0.706	0.863	0.541	1.074	0.746	0.842	0.863	18.653	23.790
10.	Chittorgarh	0.978	0.997	0.665	0.916	0.796	0.921	1.032	1.221	0.669	0.911	0.817	16.039	24.810
11.	Churu	0.93	0.68	0.000	0.597	1.046	0.686	0.427	0.840	1.876	0.787	1.022	27.871	20.100
12.	Dausa	1.077	0.876	0.079	0.528	1.051	0.746	0.470	0.650	0.838	0.702	0.752	10.312	29.050
13.	Dholpur	0.961	0.908	0.000	0.795	0.768	0.689	0.907	0.646	0.898	0.730	1.074	17.963	36.960
14.	Dungarpur	0.86	0.796	0.498	0.639	0.674	0.867	0.774	0.780	0.462	0.706	0.619	7.300	25.580
15.	Ganganagar	0.924	1.613	0.192	0.932	1.121	1.063	1.201	1.243	1.688	1.109	1.223	25.335	35.280
16.	Hanumangarh	0.876	1.602	0.202	0.795	0.906	0.778	1.025	0.845	1.707	0.971	1.164	19.996	39.330
17.	Jaipur	1.493	1.127	0.102	1.482	1.646	2.678	4.833	1.227	0.904	1.722	1.992	49.363	46.170
18.	Jaisalmer	0.79	0.946	7.763	0.421	0.958	0.850	1.179	1.066	0.709	1.631	1.135	15.031	44.540
19.	Jalore	0.845	0.871	0.022	0.561	0.526	0.715	0.716	1.009	0.680	0.661	0.741	7.591	32.180

Cont'd...

...Cont'd

20. Jhalawar	0.939	1.075	0.000	0.542	0.900	0.689	0.834	1.001	0.748	0.748	0.542	14.254	11.430
21. Jhunjhunu	0.902	1.051	1.449	0.691	1.899	1.022	0.533	0.723	1.160	1.048	0.871	20.647	21.500
22. Jodhpur	0.988	0.938	0.007	1.168	1.196	1.436	1.636	1.176	0.882	1.047	1.296	33.850	27.610
23. Karauli	0.809	0.973	0.506	0.740	0.871	0.615	0.330	0.698	0.889	0.715	0.963	14.214	35.790
24. Kota	1.916	1.073	1.332	2.266	1.394	1.847	1.960	1.571	0.999	1.595	1.913	53.463	35.860
25. Nagaur	0.847	0.838	0.432	0.595	0.877	0.740	0.539	0.661	2.931	0.940	1.095	17.201	39.240
26. Pali	0.826	0.731	1.556	1.038	0.992	0.878	0.785	1.103	0.791	0.967	0.880	21.474	20.840
27. Rajsamand	0.91	0.775	2.705	1.395	0.865	1.063	0.868	0.887	0.677	1.127	0.851	13.042	30.960
28. Sawai Madhopur	1.093	0.792	0.033	0.855	0.791	0.779	0.491	0.961	0.761	0.728	1.156	19.036	40.170
29. Sikar	0.951	1.081	0.025	0.645	1.340	0.942	0.606	0.868	1.032	0.832	0.876	20.655	21.790
30. Sirohi	0.88	1.167	3.001	2.094	0.870	1.178	1.291	1.535	0.746	1.418	0.743	17.729	18.150
31. Tonk	1.13	0.709	0.115	0.724	1.134	0.685	0.599	1.085	0.731	0.768	1.075	29.894	32.920
32. Udaipur	0.984	0.803	5.276	0.887	1.099	1.225	1.098	0.842	0.603	1.424	0.863	18.620	23.850
Average	1.00	1.00	1.00	1.00	1.00	1.00	1.00	1.00	1.00	1.00	1.00	20.69	28.87
Standard Deviation	0.21	0.25	1.71	0.54	0.31	0.41	0.78	0.25	0.52	0.30	0.35	10.97	9.67
Deviation Index	20.88	25.42	171.31	53.78	31.06	41.27	78.01	25.18	52.37	29.65	34.73	53.00	33.49

social development. Composite index of all these elements have been calculated at district level and Table 3.9 depicts the result of the analysis.

Composite index of urban process 1991–2001 (Figure 3.3) depicts the districtwise pattern of urban process. The urban process composite index is based on level of urbanisation and growth of urbanisation during 1991–2001. It becomes clear from the map that Jaipur and Kota districts are having very high degree of urbanisation with 1.99 and 1.91 index value, followed by Ajmer district with 1.39 index value. In Rajasthan 13 districts have been included in medium category in which index value is between 0.90 to 1.30. These districts are Alwar, Bharatpur, Bikaner, Churu, Ganganagar, Hanumangarh, Jodhpur, Dholpur, Jaisalmer, Nagaur, Sawai Madhopur, Tonk and Baran. Only one district of Rajasthan – Barmer is in

Figure 3.3 *Rajasthan Composite Index of Urban Process*

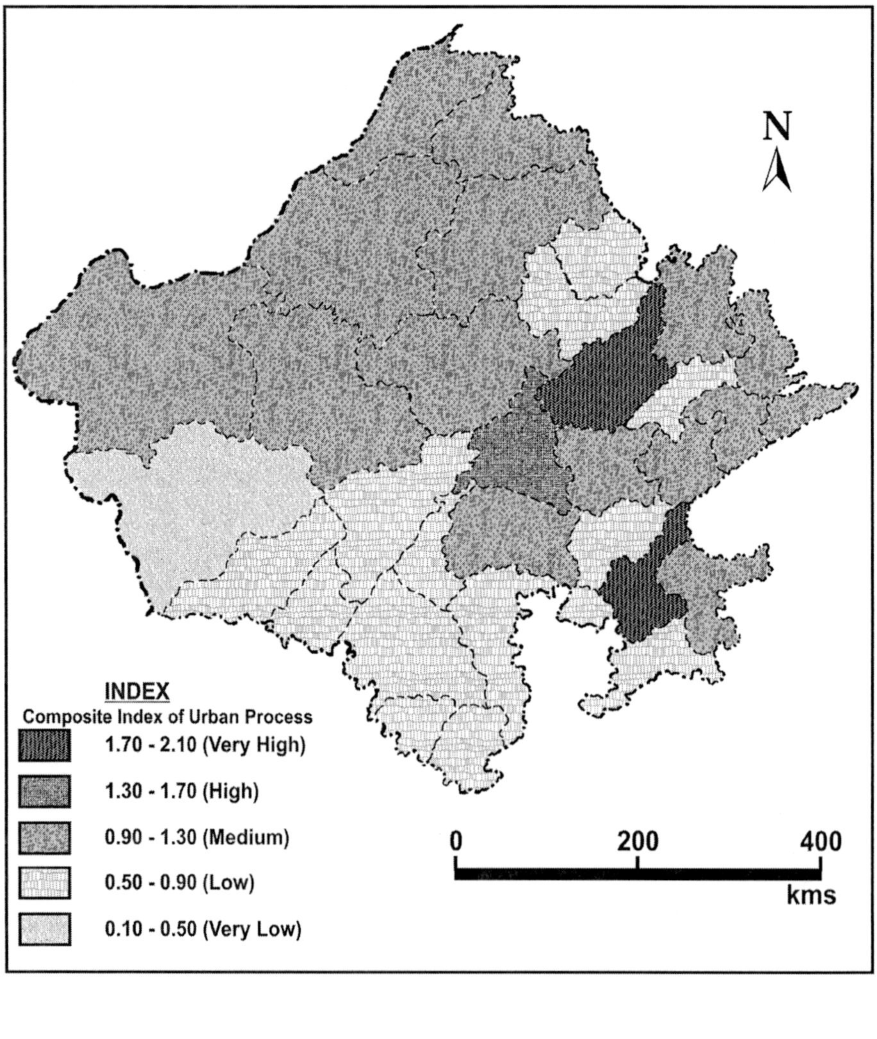

very low category with 0.19 index value. Remaining 14 districts of Rajasthan have been included in low urbanisation category in which index value varies between 0.50 to 0.90. The State average is in the category of medium.

In brief, the Rajasthan State has a long standing history of urbanisation but actual phase of urbanisation has started after independence and in 2001, the percentage of urban population was 23.38 residing in 222 towns. The main concentration of urban population is in 20 cities having more than one lakh population. The urbanisation in Rajasthan has given a new boost to the economic development and employment. But at the same time it is also a cause of degradation of environment as well as of lower quality of life in many areas of the towns. The growth oriented economic policies of the State, especially in industrial, infrastructure and tourism sector, will give a boost to urbanisation in the State in coming decades.

References

Bose, Ashish (1980), *India's Urbanisation 1901–2001*, Tata McGraw-Hill, New Delhi, p. 1.

Davis, Kingsley (1955), 'The Origin and Growth of Urbanisation in the World', *American Journal of Sociology*, March, p. 429.

Eldridge, H.T. (1956), 'The Process of Urbanisation', in J.J. Spangler and O.D. Duncan (eds), *Demographic Analysis*, Free Press, Glencoe, p. 338.

Misra, R.P. (1978), *Million Cities of India*, Vikas Publishing House Pvt. Ltd., New Delhi, p. 16.

Myrdal, Gunnar (1968), *Asian Drama*, Penguin Press, Allen Lane, London, p. 471.

Palen, J.J. (2008), *The Urban World*, Paradigm Publishers, London, pp. 8–9.

Small, J. and M. Witherick (1995), *A Modern Dictionary of Geography*, Edward Arnold, London, p. 255.

4 Degradation of Environment in Rajasthan

Environmental degradation has emerged as a major global concern for human survival at the beginning of 21st century. In the opinion of the World Commission on Environment and Development (WCED), the future generations are to face an ever-increasing environmental decay, poverty, hardships and a more polluted world. The strains of the ecological crisis are so apparent that the pressure to preserve and protect the environment has become the primary requisite of the economics of development. The resource squeeze has led to an ever-intensification of ecological damage. Land, water and air are being polluted every minute. The green revolution, has turned red. Deforestation is proceeding at an alarming rate depleting the genetic stores and mega technology has created problems of waste disposal and industrial revolution has now become a great hazard to the environment as well as to humanity.

Man's environment by now is sufficiently saturated by the complex chemical emissions, aerosols, toxic effluents, sewage, pesticides, solid wastes, polluted rains, dust and radiation. UN report on global environment reveals that the earth will become more crowded, more polluted, ecologically less stable and more vulnerable to disruption than the world we live in today. Degradation of environment has now become one of the most challenging problems not only for the nations like India but also for the states, particularly in case of cities. The environmental crisis has convened the world to use technology and resources to repair the damage already done to our environment and also to use substitutes for certain harmful chemicals in order to protect and preserve nature and natural resources. Man as a user of the environment for his developmental activities, always disrupts the natural system and creates a background for environmental degradation. When this degradation is lesser, nature recovers it by its own system of recovery, but whenever its quantity is more, it creates an imbalance in nature.

Man is dependent on the natural environment, e.g. land, water, vegetation, wildlife and atmosphere not only for his survival but also for the quality of life. The rapid urbanisation has its gratifying as well as its disquieting aspects – gratifying because of the opportunities offered for higher standard of living, education, housing, social satisfaction and public health; discrimination because of the encroachment on space, the growth of slums, the potential sprints of disease and threat to health from noise, overcrowding and the general degradation of man's physical and social environment. Pollution has been an inevitable consequence of the urbanisation and modern industrial technology, which have had direct impact on man's health and his mental, social and economic well-being.

Development (including urban development) involves exploitation of non-renewable and renewable natural resources and discharge of waste products which pollutes air, water and land. Sometimes the wastes are discharged at a rate beyond the carrying capacity of the environment and thus degrade the quality of environment. According to Paul Harrison (1975. 648), 'rape of the earth has been central to the past progress of western civilisation. The plunder and waste of irreparable resources, the poisoning and destruction of the fragile ecosystem, the belief that man can mould the nature with no need to take account of nature's reaction to such violation is a gross oversight'. Scientific and technological development has given several benefits to mankind but at the same time, it is also responsible for various types of environmental degradation. This problem is more in urban areas because of concentration of population, industries as well as of vehicles. Northam (1975) has observed two types of environmental degradations in urban areas – physical and social. 'Environmental degradation of physical nature involves contamination or impairment of use of naturally occurring elements of man's environment in which the natural element is the vehicle by which the pollutants are dispersed or also responsible for changes in atmospheric and environmental conditions'. On the other hand, 'environmental degradation of a social nature is somewhat different because it involves environmental contamination of urban places, more than of a natural element'.

In this chapter an analysis of the environmental degradation in Rajasthan has been done under two themes, first general environmental degradation in the state and second urban environmental degradation. Although, both these degradations are related with each other but from the point of view of present study, it would be better to discuss them separately.

General Pattern of Environmental Degradation in Rajasthan

Rajasthan is the largest state of India, covers an area of 3,42,274 sq. km and having 56.47 million population, residing in 222 towns and over 41,538 villages. The growing population needs more food, more houses, transportation and other infrastructural facilities. Modernisation of agriculture and industrial development through the adoption of new technology have been the means to serve this purpose. But modernisation of agriculture leads to increased use of fertilisers, agro-chemicals and pesticides, changes in land-use systems and accelerated urbanisation, while rapid industrial development leads to the production of more industrial goods, increased use of machinery, enhanced mining activities and increased facilities in the means of transport. All these tend to raise standard of living of the people and improves their quality of life. But they lead to an ecological imbalance that bears catastrophic consequences for the land. The physical environment in Rajasthan now exhibits signs of land degradation, desertification, constant droughts, deforestation, depletion of water resources, air and water pollution, and roads chocked with vehicles. Large-scale deforestation has been observed in the Aravalli mountain ranges, which has led to soil erosion, intensifying the process of desertification and the emergence of ravines and gullies.

If many of these effects are attended to the times, their severity can be brought under permissible limits in order to avoid ecological disaster in the state. It is therefore, very necessary to find ways of lessening the environmental damages and safeguarding the ecology to achieve sustainable development.

Desertification and Land Degradation

Desertification to an ecologist or environmental scientist, is a phenomenon involving deterioration of the ecosystem due to change in the pressure of human adoption or use system. It is a process through which fertile or non-desertic land is converted into dry, infertile or desertic land. According to Mann (1977), 'desertification can be defined as a package of processes which brings about certain basic changes in a particular ecosystem and converts it from a relatively non-desertic to desertic terrain. This involves an interplay of climatic, adaptive and biotic factors which may (a) work through time starting with a climatic accident and ultimately resulting in the formation of a desert, or (b) lead to accentuation of the arid conditions within already desertic terrain, or (c) cause a creep of the desertic conditions within a desertic tract towards its adjoining areas.'

In Rajasthan, arid area covers about 62 per cent of the State area and facing the problem of desertification, which is a threat not only to the environment but also to the people of this region. The Rajasthan desert is characterised by various types of sand dunes of high velocity and rolling sand, high diurnal variation of temperature, scarce rainfall (less than 300mm per annum), high wind speed, intense solar radiation and high rate of evaporation. The soils of deserts are poor in fertility, rapid infiltration rate of water, low humus content due to rapid oxidation and high salinity. Despite very hostile conditions for existence of life, this region supports large human and livestock population.

The geomorphological and archaeological evidences suggest that the desert region was once a flourishing green countryside with thick forests and well integrated drainage system of which Saraswati and Yamuna were main rivers. The study of landset data also confirms it. The onslaught of man and his domestic animals on the local ecosystem changed the panorama of the region from land of plenty to the land of poverty in less than 5,000 years.

In this part of Rajasthan due to high pressure of grazing lands and overexploitation of resources, degradation of environment is growing day by day. Increased desertification problems are also evident in the form of salinity hazard, seepage and waterlogging problems of the desert. The increased pressure of livestock results in overexploitation of resources leading to soil erosion and removal of surface vegetation. In fact, desertification is a very serious problem of Rajasthan, for which measures of control should be adopted. The following steps should be taken for the control of desertification:

(i) Ban on cutting of desert vegetation,
(ii) Plantation of ecologically suitable plants,
(iii) Development of pasture lands,
(iv) Control of overgrazing,
(v) Stabilisation of sand dunes,
(vi) Proper use of available ground water resources,
(vii) Management of land use and farming practices, and
(viii) Public awareness should be generated.

The problem of land degradation in the form of ravine and gully erosion is a major menace to agriculture in Rajasthan. Kota, Bundi, Sawai Madhopur, Karauli and Dholpur districts account for 77 per cent of the total area affected by ravines in Rajasthan. Ravines have developed along river Chambal, Kali Sindh, Banas, Banganga, Sahibi and Mahi rivers in districts of Kota, Bundi, Jhalawar, Tonk, Sawai Madhopur, Karauli, Alwar, Banswara and Durgapur. Efforts have been made in the recent past to reclaim ravines, but the success rate has been low.

Drought

Drought concerns with a particular state of an area when rainfall deficiency has reached to such an extent that the moisture available in the soil may not be found sufficient for normal growth of plant and trees. An analysis of Rajasthan's rainfall data indicates that 40 out of the last 50 years have been drought years. The analysis further reveals that the climate of Rajasthan has exhibited a continuing trend towards desiccation, particularly after the severe drought of 1987. Rainfall has been substantially lower in the last three decades of the 20th century.

In Rajasthan, droughts are caused mainly by the failure of rainfall from south-west monsoon. Recent studies on global circulations show that El-Nino phase of the Southern Oscillations (ENSO) has the largest impact on India by way of droughts. Besides, climatic changes as a result of increased concentration of the atmospheric carbon dioxide, methane and nitrous oxide also influence the frequency of droughts in the region. The causes of drought in Rajasthan are region's geographic location not favouring abundant monsoon rainfall, poor quality and deep ground water, absence of perennial rivers and forests, the poor water holding capacity of the soil, and huge withdrawal of limited ground water resources. The Rajasthan State is a highly drought prone state and nearly two-third of its area is usually affected by drought on alternative year. In spite of several measures taken by the government, drought situation may continue in Rajasthan and it is a threat not only to the people and economy of the State but also to the environment.

Deforestation

The herculean feats of economic gains and development have been at the cost of deforestation. Deforestation is one of the outcomes of various developmental activities having the character of extensional development and it is true that deforestation is a havoc wreaked by thoughtless destructive activities of development. Forest clearings were made not only for village settlements but also for cultivation and pastures. As the population increased more forests were cleared for various uses. Apart from this, the commercial exploitation of forests is the main cause of deforestation.

The forest area of Rajasthan in year 2005–06 was 23,627 sq. km, which is 9.54 per cent of the total geographical area of the State. The percentage of forest areas varies from 0.42 per cent in Churu district to 38.59 per cent in Karauli district. The district having less than 5 per cent forest area are Jaisalmer, Hanumangarh, Jodhpur, Nagaur, Bikaner, Ganganagar, Tonk, Barmer and Jalore. Although, the total forest area has increased 2.75 per cent during last two decades but it is still much less than required average. Efforts have been made in the past to ensure people's participation in forest by promoting social forestry, agro-forestry and silviculture, but these efforts failed to develop the ecology of the State to any meaningful extent.

Scarcity of Water

Rajasthan has 1.2 per cent of India's water resources in the form of annual rainfall, against a more than 10.4 per cent share in geographical area. This means that water is the single most scarce resource in the state. The scarcity of water in Rajasthan has become increasingly acute in the last two decades. The position of ground water is not satisfactory in the State because of overexploitation. Rajasthan has 33 districts and 237 blocks. In 2001, 49 blocks were classified in the 'safe' category, 21 in 'semi-critical', 80 in 'critical', 86 in 'overexploited' and one in the saline category. The block with saline ground water is Taranagar in Churu district. Quite a large number of the overexploited blocks fall in the Alwar, Jaipur, Jalore, Jhunjhunu and

Nagaur districts. The people of these villages are not assured of having safe drinking water supply. The water level of underground water is going down and down, thus, a matter of serious concern not only for the people but also for the ecosystem of the state.

The areas with poor quality of ground water are as follows:

Urban agglomerates	:	Jaipur, Jodhpur, Udaipur
Industrial pollution	:	Bhilwara, Udaipur, Jaipur (Sanganer), Kota, Pali, Barmer (Balotra), Jodhpur, etc.
Base metal mining and processing	:	Zawar, Agucha, Rajpura, Khetri, Bichri
Thermal power generation	:	Kota, Suratgarh
Marble mining and polishing	:	Makrana, Rajsamand, Andhi, Suket, Jhiri, Kishangarh

The Fluoride Problem

Another problem related with ground water resources of Rajasthan is the presence of fluoride, which is very harmful to the health of the people. An estimate shows that more than 6,000 villages of Rajasthan are suffering from the scourge of fluorosis, caused by excessive fluoride in drinking water which leads to degeneration of teeth and humped backs. Fluorosis is both widespread and acute in Rajasthan. Nearly 3.5 lakh inhabitants of the desert belt of Rajasthan suffer from fluorosis. An entire sub-region called 'Banka Patti' (humpback strip) is in Nagaur district, where most of the villagers suffer from fluorosis. The worst fluoride affected districts of the State are Jaipur, Nagaur, Sikar, Jodhpur, Barmer, Ajmer, Sirohi, Jhunjhunu, Churu, Ganganagar and Tonk.

Salinity and Waterlogging

In canal irrigated areas of Rajasthan, the problem of salinity and waterlogging is also a cause of environmental degradation. Improper water management of canal water leads to salinity and waterlogging. Salinity of soil affects the agricultural productivity of land. In many desert districts of Rajasthan salinity is of natural type, comprising of '*ranns*' and other saline depressions. These depressions are affected by severe to very severe salinity, while the palaeochannels are afflicted by severe salinity. On the other hand man-induced salinity has resulted in recent times because of the construction of tanks and canals along the courses of the earlier drainage channels. It is also due to over irrigation, where sub-soil salts came up and deposited on soil. Such problem is common in Lunkaransar area of Bikaner district and also in its adjoining areas.

Waterlogging conditions usually occur in areas where ground water recharge is in excess of ground water discharge. In such areas, the ground water storage level starts rising, and this is reflected in the use of ground water canals. As the water table reaches within two metres of the land surface, the water table starts losing excess water through evapotranspiration, either through direct contact with the atmosphere via a capillary zone or through perennial vegetation and crops.

In Rajasthan waterlogging conditions are evident in parts of Kota and Bundi districts because of the surface irrigation in Chambal command area. But in this part waterlogging is limited. On the other hand large areas of waterlogging can be seen in Indira Gandhi Canal irrigated areas of Suratgarh, Anupgarh, Sripurwala, Badopal, Ravatsar and Rangmahal. In these areas not only crops and vegetation have been destroyed but several villages have been deserted and the whole area of Badopal has become just like a lake. Such position may arise in other canal irrigated areas. Due to waterlogging the whole eco system has been adversely affected and it has become a main cause of environmental degradation in such areas. To control such problems necessary technical steps should be taken by the concerned authorities.

Loss of Biodiversity and Wildlife

The loss of biodiversity and wildlife is also a cause of imbalance in ecosystem. Most of the endangered species of birds and mammals in Rajasthan are on the brink of extinction. The population of eagles, storks, pochards, tigers, great Indian bustard and many other species has shrunk to remote habitats in the State. The constant onslaught on them by humans, coupled with their inbreeding tendencies, may wipe them out totally. Rajasthan has two national parks and 23 wildlife sanctuaries for the protection of wildlife and biodiversity, but in spite of that Sariska Game Sanctuary which was meant especially to protect tigers, now has only a couple of tigers, that too relocated from Ranthambore sanctuary.

The plants most at risk of extinction in Rajasthan are *Phog, Rohida, Khejri, Babool, Kumat, Jal, Sevan, Ker, Nagori, Ashgand* and *Guggal*. Among the animals, the *Chinkara, Bustard* and *Black Buck* are already endangered species. The degree of endemism and danger for plants and animal species in the desert of Rajasthan is comparatively higher than in other deserts of the world. The conservation of biological diversity requires a major shift at all levels of decision making and also needs public participation.

In brief, following general problems of environmental degradation in Rajasthan need priority consideration in the planning:

(i) Recurring drought and acceleration of desertification process,
(ii) Destruction of natural vegetation either by deforestation or by overgrazing,
(iii) Scarcity of water, especially of drinking water,
(iv) Problem of land degradation,
(v) Salinity hazards, waterlogging in the canal command areas,

(vi) Occurrence of brackish ground water,
(vii) Creation of scarred and derelict landscape due to open cast mining of minerals and raw material resources, and
(viii) Depletion of biodiversity and stress on wildlife habitat.

Urban Environmental Degradation

Environmental degradation in urban areas is more in comparison to other areas. This is more true in case of towns/cities having more than one lakh population. The general problems of environmental degradation discussed above in this chapter are also common to urban areas. But urban areas are facing many other serious problems of environmental degradation. One of the objective of the present study is to analyse the specific problems of the urban environmental degradation in context with cities of above one lakh population.

At the beginning of 21st century, city systems have become a dominant factor in Rajasthan's social, economic, cultural and political matrix. Burdened with all the problems of economic growth, the State's cities are increasingly subject to dramatic crises, such as unemployment, environmental degradation, lack of urban civic services, deterioration of existing infrastructure and lack of access to land, finance and adequate shelter. The cities will increasingly become test beds for the adequacy of political institutions, the better performance of government agencies, and the effectiveness of programmes to combat social exclusion, protect and repair the environment and promote human development in a meaningful manner.

Like most other states, Rajasthan also faces problem of unchecked urbanisation. Out of 222 towns and cities in the State, 20 cities have attained more than one lakh population. According to 2001 census Jaipur, the capital city, is the highest one having total population of 23,24,319 with highest growth rate of 59.37 per cent during 1991–2001. The cities having more than five lakh population are Jodhpur, Kota and Bikaner. While Ajmer, Udaipur, Bhilwara, Alwar, Sriganganagar and Bharatpur have recorded population between two to five lakh; while other ten cities are having population between one to two lakh in 2001. All these 20 cities of the State having above one lakh population are now facing problems of environmental degradation. The higher the population of the city more are the problems of environmental degradation. The important problems of urban environmental degradation are as follows:

Lack of Sanitation Facilities

One of the glaring problem that cities and towns in Rajasthan confront is the lack of sanitation, which is also a main cause of urban environmental degradation. The prevailing conditions in cities of the State are indicative of gross neglect of sanitation services, not just over last few years, but over several decades. Most of the cities are not having proper sewerage system and even where these system do exist, their capacities are not enough to cope with the increasing pressure being put on them. In Rajasthan, top ten cities, e.g. Jaipur, Jodhpur, Kota, Bikaner, Ajmer, Udaipur, Bhilwara, Alwar, Sriganganagar and Bharatpur are facing sewerage problems, the underground system is limited and only except Jaipur, other cities are not having sewerage treatment plants of proper capacity. For example Kota, one of the prime city of Rajasthan is yet to develop a proper underground sewerage system. In most of the cities the waste water drains are open and not cleaned properly, therefore, become a main cause of pollution. In 2001, urban population of 5.8 million, now more than 9 million is not having access to proper sanitation facilities. In every city, slums have been developed, which are not having minimum sanitation facilities. These areas are disease prone areas and facing all types of environmental degradation.

Absence of a proper sewerage system in the newly developed and developing housing colonies, is one of the major problems of sanitation. It not only leads to deterioration of the quality of areas, but also a root cause of numerous infectious and parasitic diseases. Lack of paved roads in many of these residential colonies leads to accumulation of water and slush during rainy season, providing ideal conditions for breeding mosquitoes. These puddles of water and heaps of garbage along the roads facilitate the proliferation, vectors including mosquitoes, houseflies, various arthropods and helminths. The drainage channels in cities are usually clogged with waste and industrial effluents, and reduced to sluggishly flowing cesspools, which pose immense health hazards. Stray animals, including cattle, pigs and dogs, which form a part and parcel of cities, are another source of insanitary conditions, resulting in further health risks. In most of the cities of Rajasthan there is a lack of proper system of waste water disposal in housing units. More than 50 per cent of urban houses have a system of direct pipes to roadside drain. The rest dispose their waste water either onto open lands or streets without drains. A number of problems arise due to this, not the least of them being unhygienic environment around houses.

Problem of Urban Waste Disposal

All the urban centres including cities and towns of Rajasthan are facing the waste disposal problem. The urban waste now has become one of the main cause of the degradation of urban environment. Urban wastes are discarded organic or inorganic substance in the form of solid, semi-solid, liquid and gases which are residues or derivatives of human, vegetable materials and industry. Urban wastes are broadly classified as under:

(i) *Household Wastes*: Consisting of human excreta, waste created in preparation of food, generally in terms of garbage, ash, etc.

(ii) *City Wastes*: All the matter derived from the maintenance of streets, roads, parks and schools, paper, dry leaves, animal wastes, sludge, carcasses of small animals, street and drain mud, slaughter house wastes, etc.

(iii) *Commercial Wastes*: Resulting from activity in office buildings, stores, markets, theatres, hospitals and restaurants, etc.

(iv) *Industrial Wastes*: Iron and Ferro alloy ores shipped to mills, other metal ore processed, construction materials and non-metals including lime, minerals, fuels, used petrochemical feed stocks and other non-fuels purposes, e.g. asphalt, tar, lubricant and wastes.

(v) *Liquid Wastes*: Domestic sewage and factory effluents.

(vi) *Gaseous Wastes*: Gases discharged by above mentioned wastes

According to an estimate the output of urban refuse amounts to about 0.3 kg of *kachra* (household-refuse, city wastes and commercial wastes), 0.2 kg of night soil and 0.8 kg liquid waste per head per day. There is however considerable variation from city to city, depending upon local factors such as industrial activities, volume of commercial activities, dilution of night soil, ablution waste, etc. It is estimated that an average urban waste in a million city is about 33,800 tonnes daily. But in Rajasthan the quantum of urban waste is lesser because the population size of the urban centres (excluding Jaipur) is quite lower. As per rough estimates, 400 tonnes of solid waste is generated per day in Jaipur city – 3.8 kg per day per person. As per another estimates, the generation of liquid waste in Jaipur per day is about 283 kilolitres. In Jaipur city, according to a rough estimate 1,38,000 kg per month hospital waste is generated. The amount of hazardous household waste and waste from small industries and medical clinics is increasing day by day, which requires proper waste management.

The study of Bikaner shows an average per capita generation of solid waste in the lower class neighbourhood was 76 gm/capita/day, while it was 249 gm/capita/day in higher class neighbourhoods. The study was conducted in three upper class neighbourhoods, e.g. Jawar Nagar, Kailash Puri and Murlidhar Vyas Nagar and three lower class neighbourhoods of Harijanon ki Bari, Guwar, Rampura Basti and Shiv Bari. The difference between higher and lower class is obvious due to the variable buying capacity and consumption pattern in two classes of neighbourhoods. The proportion of organic portion in solid wastes of lower class households was higher (51.7%) than in wastes of upper class households (45.7%). This is because the people living in the lower class areas have to spend a bigger proportion of their meagre income on buying the basic necessity of food itself. On the other hand, the richer people have to spend a lower proportion of their comparatively larger income on food and they spend more on consumable items made of paper, plastics and metals. There is a spatial variation in quantity and composition of solid wastes generated in different areas of the city but in absence of proper management system the waste become a cause of environmental degradation. In terms of collection and disposal of waste only 25 per cent of the total solid waste is properly disposed daily, remaining 75 per cent collected either twice or once in a month and its disposal is also not proper. According to sanitation department of Bikaner, only 27 out of the 55 wards of the city are being provided waste collection and disposal services. The only earmarked site for waste dumping is Gogagate waste dumping ground. This waste dumping site is not a sanitary landfill site, the leaches from the dumping site are polluting the ground water of the surrounding water and also causing land pollution.

Kota city is also having a great problem of waste collection and disposal. During last two decades Kota has expanded at a fast rate but its system of waste collection has not improved much. The municipal council of the city is making necessary arrangement for waste collection but not able to collect the solid waste, what to say about the disposal of waste. The colonies of the city which are worst in terms of waste collection are – Ram Nagar, Ganeshpal, Balakund, Durga Basti, Shivpura, Bavdi, Indra Colony, Surya Nagar, Near D.C.M. Circle, Raipura, Thikra, Bajrang Nagar, Sanjay Nagar A&B, Adarsh Nagar, Dostpura A&B, Brijraj Colony, Adarshila, Balakund, Oriya Basti (Dadabari and Talwandi), Prem Nagar, Govind Nagar, Ghodewala Baba, along river Chambal and Chhatarpur. In old Kota city also proper arrangement of regular collection of waste has not been done.

The conditions of other towns of Rajasthan are also more or less similar in terms of waste collection and disposal. The small municipal boards are not having even financial resources for the proper management of waste collection and disposal. Of all the wastes (municipal, industrial, construction and demolition) the municipal solid waste (MSW) are the most non-homogenous since they consist of the residues of nearly all materials used by people such as organic wastes, papers, plastics, fabrics, leather, metals, glass and miscellaneous other materials. This heterogeneous waste makes it difficult to treat it with one effective method. Only 5 to 10 per cent of the solid waste gets treatment before disposal, a huge chunk of the collected waste gets dumped at the landfills where it rots and emits health hazardous methane gas. In most of the cities and towns, waste treatment plants have not been established, even where exists their capacity is not according to load. The limited financial resources and poor technical means do not leave enough scope for municipal councils/boards or similar agencies to plan for effectively meeting out the ever-increasing demands on them.

Industrial Pollution

Industries, which are considered as a symbol of development, are also a major cause of pollution. Since industries and industrial areas are mostly developed in urban areas they have become a cause of environmental degradation. The nature of industrial pollution is multifacial because industries used to create air, water, land as well as noise pollution. In present study industrial pollution has been studied in selected towns of Rajasthan. In this context, a brief description of industrial pollution at Kota, Jodhpur, Pali, Alwar and Bhilwara is given here in order to explain the nature of pollution.

Kota

Kota is among the big cities of Rajasthan and also known as an industrial city. The development of various types of industries have given it a special status but at the same time these industries are responsible for the degradation of urban environment. At present there are 15 large- and medium-scale industries and the number of small-scale industries is more than 6,000 employing over one lakh people. The main industrial areas of Kota are Kansuwa Industrial Area, Anantpura Industrial Area, Indraprastha Industrial Area, Ranpur Industrial Area, Bhadana Industrial Area and other industrial areas located at Baran Road, Sakatpura, etc. The main industries located at Kota are K.T.P.S. Kota, C.F.C.L., Shriram Rayons and Fertiliser, Instrumentation Ltd., Om Metals, Infra Project Ltd., Samcor Glass Ltd., Tilam Sangh, Rajasthan Kota Project, Multimetals Ltd., Kota Zila Dugdh Utpadak Sahkari Sangh Ltd., Nobal Grain India Pvt. Ltd. and Injocom India Pvt. Ltd.

The industrial areas around and in Kota city have had direct impact on environment as well as on people. The Kansuwa industrial area is the major source of pollution. The existing industries are the main source of air pollution. The existing level of pollution during winter period has been worked out by the Rajasthan State Pollution Board considering five sample points. The result showed particulates matter concentration of 457.92 ug/m^3 at the commercial site, 324.61 ug/m^3 at industrial sites and 224.36 ug/m^3 at a comparatively cleaner site. From these results, it can be formulated that the presence of significant amount of particulates matter in the ambient atmosphere (about 230 ug/m^3) is due to industrial and commercial activities. Besides particulates matter, the oxides of nitrogen are also of great concern. The air pollution caused by the industries located in the Kansuwa area have a direct impact on city because winds flow from south-east direction used to create pollution. The existence of thermal power plant is also responsible for the air pollution in large part of the city. The air pollution problem by coal fired in thermal power plant is directly related to the combustion of coal ingredients. The chief pollutants generated during coal combustion are CO, CO_2, SO_2, Nitrogen Oxide and fly ash. The location of thermal power plant is such that winds blowing from north-west direction affect two-third of the city. The percentage frequency of winds blowing from this direction is 40.89 during summer and 22.82 during winter, thus leaving adverse effects not only on the people but also on vegetation and water bodies.

Industries are also responsible for water pollution. Out of 17 big industrial units 13 units are discharging more than 50,000 litres waste water per day. Most of this waste water ultimately goes to river Chambal. The composition of the waste water generated is quite complex in nature and difficult to treat. The pollutants like residual chlorine, ammonia, BOD, COD, trace metals, mercury, dissolved solids, etc. or part of the waste water. This waste water used to damage the biotic system of Chambal river and is also harmful when the water is used for drinking purpose both for human and animals.

Jodhpur

There are six industrial areas at Jodhpur city, namely Marudhar Industrial Area, Barni, Phase–I and Phase–II, Light Industrial Area, Heavy Industrial Area, Industrial Area behind Power House and Industrial Estate. The industries of Jodhpur are the main cause of water pollution. The total effluent generated from various industrial areas is about 4.5 million litre per day. The dyeing and printing industries of the Jodhpur city are mostly concentrated in Basni Phase–II industrial area. This area consists of 68.75 per cent of the dyeing and printing units of the city. The contaminated and polluted water is carried through a large drain in the south-east of the industrial area. Although CEPT plant has been established with a capacity of 5 MLD to treat the polluted water but it is insufficient, hence most of the waste water is drained untreated into the dry bed of the Jojri river flowing south-east of the industrial area. In dyeing and printing use of synthetic chemical has affected the ground water. The chemicals used in this industry include bleaching agents, caustic dye, synthetics, sodium silicate, starch gum, etc. The waste water quantity used in the printing varies from 100 to 500 litre per 100 m length of cloth. The waste water discharged from these industries contain BOD, COD and TDS. The effluents emerging from silicate washing and washing of guar gums, starch, etc. are also contributing to pollution significantly. Apart from water pollution, air pollution to a limited extent is also there but its impact is localised.

Pali

Pali town is famous for the dying and printing of cotton fabrics. The industrial complexes presently accommodate about 767 industrial units, majority of them are in notified industrial areas of Phase–I, Phase–II and Phase–III. Only 37 per cent of the industrial units are scattered in different parts of Pali city. Industrial waste water from about 363 textile units situated in Mandia road industrial area Phase–III is being treated at existing effluent treatment plant at Mandia road on the bank of Bandi river. The treated effluent is discharged into the river. The untreated industrial waste water from other industrial units spread at various locations in and around the city and it is being discharged in the river at different plants and it is also a cause of pollution.

River Bandi before entering the Pali town, approximately 3 km of the stream is completely dry. The drains and industrial as well as domestic waste water enter into the river from different points. After receiving this waste water the river flows approximately 35 km down to Nehru Dam. Due to continuous waste water discharges from textile units in last 30 years this waste water has seeped into underground water strata and the wells located within 100 m on the banks of both sides of the river are contaminated up to 35 km. These wells have been so polluted that they are now unfit for drinking and irrigation purposes.

Alwar

The main industrial area of Alwar is the Matsya Industrial Area. The old industrial area is located near railway station, now surrounded by residential area. The Matsya industrial area is having 462 industrial plots and is one of the biggest industrial areas of the state in which 283 industrial units are working. It has 15 large-scale, 18 medium scale and 250 small-scale units, which include chemical, engineering, steel, textile, plastic, paper and food processing units. From Matsya and old industrial area of Alwar (MLD) industrial effluent is being discharged by 174 water polluting industries, which is directly flowing towards Hans Sarovar. It has polluted the water of the Hans Sarovar. There are several chemical industrial units located in Matsya industrial area. Effluent is generated from acid coolers, water treatment plant, scrubber contains pH, COD, TSS, phosphates and fluorides. The industrial units located at old industrial area are also responsible for air pollution in the city.

Bhilwara

Bhilwara is also an industrial town of Rajasthan and famous for its textile and synthetic cloth manufacturing. It is also known as '*Textile City*' of Rajasthan. The rapid growth of industries in and around Bhilwara town now has became a cause of industrial pollution. Industrial plants, where boilers and electric generating machines are located, are the big source of air pollution. In addition to this, sulphur fumes also contribute to air pollution around dyeing plants. The location of the industries around the city is not at all ideal, because city is located at the windward side of the industrial area and the industrial smoke flows towards the city causing air pollution.

At Bhilwara out of total industrial effluents that are discharged by industries 85 per cent flows via Guwardi nallah into river Banas and about 14 per cent goes into the Kothari river. The main pollutants occurring in the combined industrial effluent are as follows:

Total Suspended Solids (TSS) 30–250
Chemical Oxygen Demand (COD) up to 90 mg/lt.
Bio Chemical Oxygen Demand (BOD) 20–45 mg/lt.
Total Dissolved Solids (TDS) 3,500–6,500 mg/lt.

The industrial effluent flows into the natural water courses. The seepage of waste water into sub-soil is a cause of sub-soil water pollution. Most of the industries are not following the required effluent discharge standards. Due to discharge of polluted waste water by these units ground water is being affected adversely.

Examples of the above mentioned cities clearly indicate that industrial pollution is common in all the cities having above one lakh population. The nature and degree of pollution varies from city to city according to the nature and pattern of industries.

Urban Air Pollution by Automobiles and Noise Pollution

The main cause of urban air pollution is the automobiles. With the increase of urbanisation and rapid growth of population, demand for urban traffic has increased many times. In last two decades, there has been a tremendous growth of metorised vehicles in cities of Rajasthan. This increase is more in big cities of the State, e.g. Jaipur, Jodhpur, Kota, Bikaner, Ajmer, Udaipur, Bhilwara, Alwar and Sriganganagar causing several problems like air pollution, noise pollution, parking problems, traffic congestion and increase in accidents.

Automobiles are considered as the main source of air pollution. Automobiles exhaust release smoke and a certain extent of lead particle. Smoke contains the gaseous pollutants like carbon monoxide, hydrocarbons and oxides of nitrogen. The pollutants emitted by petrol and diesel driven vehicles are given in Table 4.1.

Table 4.1 *Pollutants Emitted by Petrol and Diesel Driven Vehicles*

Emissions	*Petrol Engine*	*Diesel Engine*
Carbon Monoxide	3	–
Sulphur Dioxide	40 ppm	400 ppm
Nitrogen Oxide	1100 ppm	300 ppm
Unburnt and Hydrocarbons	150 ppm	20 ppm
Particulates Matter	0.1 g/m3	0.01 g/m3

Air pollution in urban areas is increasing because of heavy traffic contributing hydrocarbons, carbon monoxide, suspended particulate

(SPM) and sulphur dioxide. According to an estimate, the level of SPM in heavy traffic urban areas of Rajasthan was recorded 700–800 mg per metre cube, which is much higher than the permissible level of 500 mg per metre cube. Sulphur dioxide levels are also very high at 60–70 mg per metre cube against the permissible limit of 40 mg per metre cube. For Rajasthan, as a whole, the vehicular air pollution load has been estimated to the tune of 2,99,260 tonnes a year. An example of Kota city clearly indicates the vehicular emission and their amount at various sites (Table 4.2).

Table 4.2 *Automobiles Emission in Kota City (March 2008)*

Area	*CO2 (%)*	*NO2 (%)*	*H2 (%)*	*Particular (mg/m3)*
Industrial Area	60	25	20	200 – 650
Sriram Nagar	60	20	20	160 – 500
Dadabari	66	25	30	–
CAD Circle	60	21	27	–
Vigyan Nagar Circle	61	27	30	–
Aerodrome Circle	70	25	35	–
Gumanpura	67	16	27	–
Kotri Circle	64	12	22	–
Rampura	65	18	30	190 – 188
Nayapura	66	14	26	100 – 300
Antaghar	65	13	25	–

In Kota city there are more than 2,50,000 vehicles and their number is increasing day by day causing a great problem of urban air pollution.

Road traffic is the main source of noise pollution, which contributes about 55 to 60 per cent of the total city noise pollution in every locality in Jaipur, noise pollution levels exceed (at 78–87 decibels) the permissible level of 70 decibels. This is more true about commercial zone which includes Choti Chopper, Badi Chopper, Ajmeri Gate, Tonk Phatak, Railway Station, Raja Park where noise level varies between 74 to 82 db. Even in silence zone of SMS Hospital, SDMH and High Court area the recorded noise level at day time was 69.7 to 77.2 db. As the city is growing noise level is also increasing. This fact is also true about Jodhpur, Kota, Bikaner, Udaipur and Ajmer.

The spatial pattern of air, water, noise and land pollution in sample towns of Rajasthan has been discussed in Chapter 5.

References

Mann, H.S. (1977), 'Desertification: An Overview', *Arid Zone*, 16 (3), pp. 1–4.

Northam, R.M. (1975), *Urban Geography*, John Wiley & Sons, New York, p. 76.

5 Spatial Pattern of Pollution in Urban Areas

The urban environment is an interaction between physical, biological and socio-cultural elements that are in a dynamic process of transformation. The phenomenon of urbanisation is man-made, being only indirectly influenced by natural conditions. Hence, the characteristics of the urban environment are largely human creations that fashion the life support system and bring about direct and indirect hazards that pose a challenge to man's health and well-being (Giggs, 1979; Eyles, 1982). The cultural environment is controlled by the human being such as houses, roads, industries, etc. Nowadays, with the increase of population in urban areas the pollution of environment has increased substantially. With the continuing growth, urban areas are becoming the major source of green house gases from vehicles, industries and buildings. In addition, cities attract and consume vast amounts of water, metals, fuels, paper, glass, plastics and building materials, and then produce waste water, sewage sludge, local air pollution, demolition waste, and municipal waste. Thus, urban centres now become areas where pollution is maximum.

The problem of pollution in urban areas is growing day by day, which not only degrades the environment but also have/had adverse effect on human health and quality of life. In Chapter 4, some general problems of environmental degradation in urban areas have been discussed. The main problem of urban areas is pollution, i.e. air, water, land and noise pollution, which needs special attention not only for the understanding of present position of urban pollution but also for its planning. Although, in most of the cases whole of the city is under the grip of pollution, but there are specific areas which are highly polluted, while other areas have either moderate or minimum pollution. Therefore, there is a need to discuss the spatial pattern of air, water, land and noise pollution in sample cities. The samples selected for the study are as follows:

1. Kota
2. Bharatpur
3. Sriganganagar
4. Tonk, and
5. Gangapur City

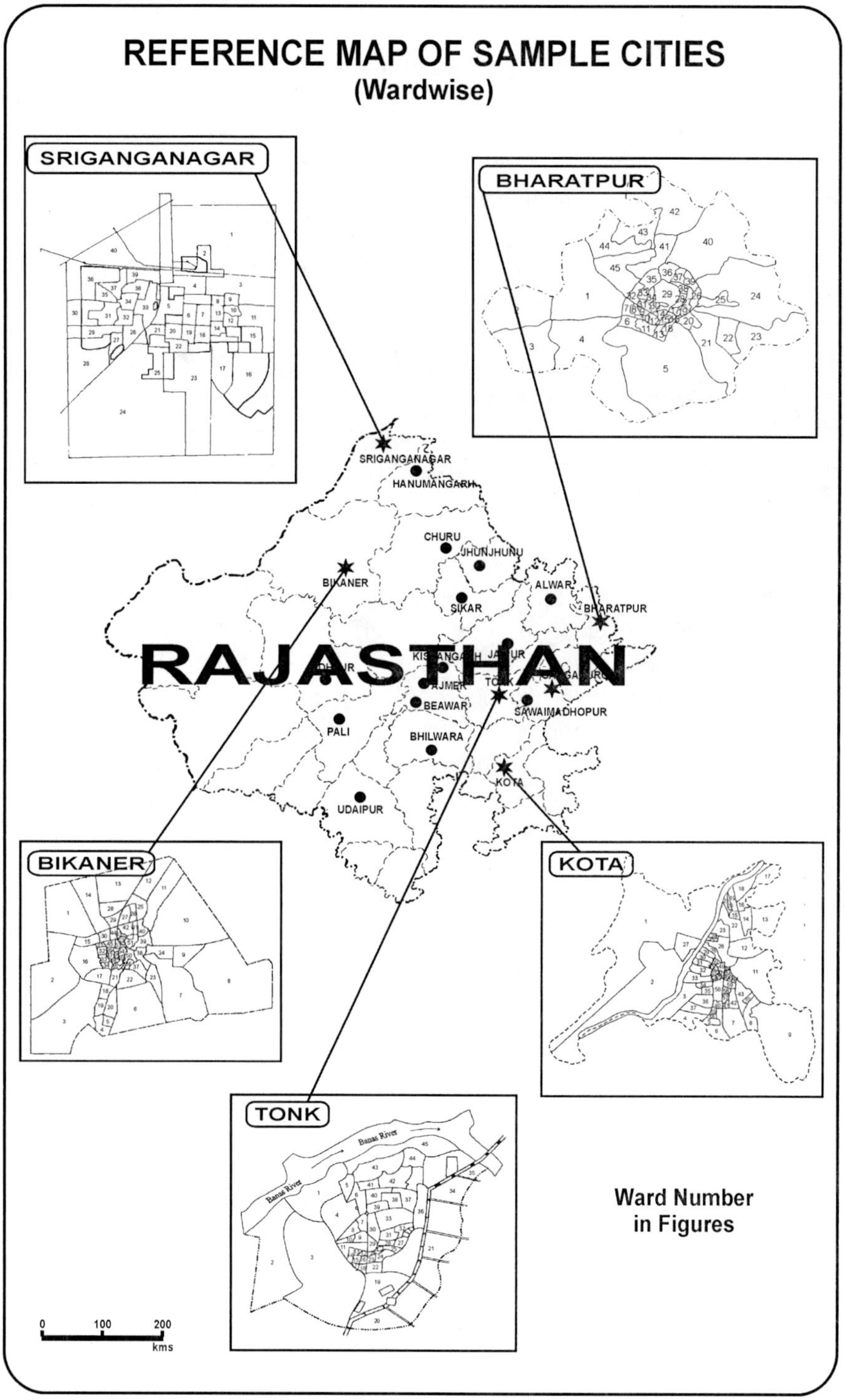
REFERENCE MAP OF SAMPLE CITIES
(Wardwise)
SRIGANGANAGAR
BHARATPUR
BIKANER
KOTA
TONK
RAJASTHAN
HANUMANGARH
CHURU
JHUNJHUNU
ALWAR
SIKAR
AJMER
BEAWAR
SAWAIMADHOPUR
PALI
BHILWARA
UDAIPUR
Banas River
Ward Number in Figures
0 100 200
kms

Kota

Kota (78°10' E; 23°11' N) is situated in the south-eastern part of Rajasthan known as Hadoti Plateau, it ranks third in terms of population of cities in Rajasthan. According to 2001 census the total population of Kota city was 7,04,731 and it is next to Jaipur and Jodhpur. Kota is an industrial as well as commercial town of Rajasthan and during the last two decades it is also known for coaching classes. The reference of coaching is necessary because more than 50,000 students are studying here, which have a direct impact not only on the economy of the city but also on the quality of life as well as pollution pattern of the city.

The study of the spatial pattern of pollution of Kota city, in general reflects the pollution patterns of big cities in Rajasthan. A detailed study of Kota city has been done for the identification of areas having high, medium and low pollution. The study includes air, water, noise and land pollution and for each type of pollution maps have been prepared. The data available from State Pollution Board have also been utilised for analysis.

Air Pollution

Air pollution is the result of combined effects of several pollutants. The industrial, automotive and domestic activities have resulted in increasingly outrageous insults to the atmosphere. In air pollution the pollutants include gasses and solids and liquid particles of both organic and inorganic chemical classification (Strahler and Strahler, 1977). The main reasons for the increasing rate of air pollution in cities like Kota are:

(i) Poisonous gases and other particles emitted from industries without treatment;
(ii) Increase in the number of automobiles and their emissions like carbon monoxide, sulphur dioxide, nitrogen oxides, unburnt hydrocarbons and particulate matter;
(iii) Increased use of chemicals and petrochemicals, and
(iv) Population concentration.

The spatial pattern of air pollution, i.e. air pollution areas of Kota city has been depicted in Figure 5.1. It becomes clear from the map that there are three areas of Kota having high air pollution. These are (i) industrial area of south-east Kota known as Indraprasth Industrial Area, (ii) area adjoining to Super Thermal Power Plant, and (iii) railway area of Bhimganj Mandi. Other areas of air pollution are the traffic points located on main roads of the city.

The industrial area of Kota located in south-eastern part of the city is a highly air polluted area. The pattern of pollution and its nature is shown in Table 5.1 and Figure 5.2.

Figure 5.1 *Kota: Air Polluted Areas*

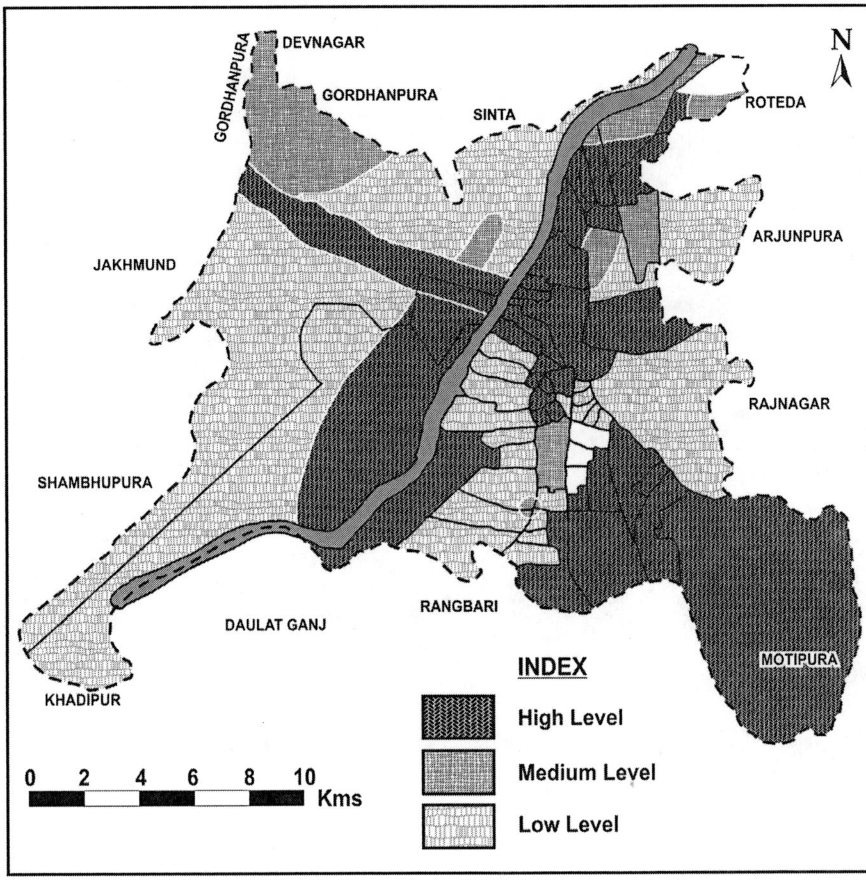

Table 5.1 *Air Pollution in Industrial Area of Kota City*

Year		SO_2	NO_2	*RSPM*	*SPM*
2007	Maximum	15.1	58.4	276	529
	Minimum	5.3	9.8	14	30
	Standard Deviation	2.9	12.5	70	147
	Mean	9	20.1	102	232
	No. of Days	52	52	52	52
2008	Maximum	23.9	52.4	348	568
	Minimum	5.7	11.4	36	68
	Standard Deviation	4	10.6	65	125
	Mean	10.7	27.2	144	318
	No. of Days	61	61	61	61

Source: www.cpcbedb.nic.in

Figure 5.2 *Air Pollution in Indraprasth Industrial Area, Kota*

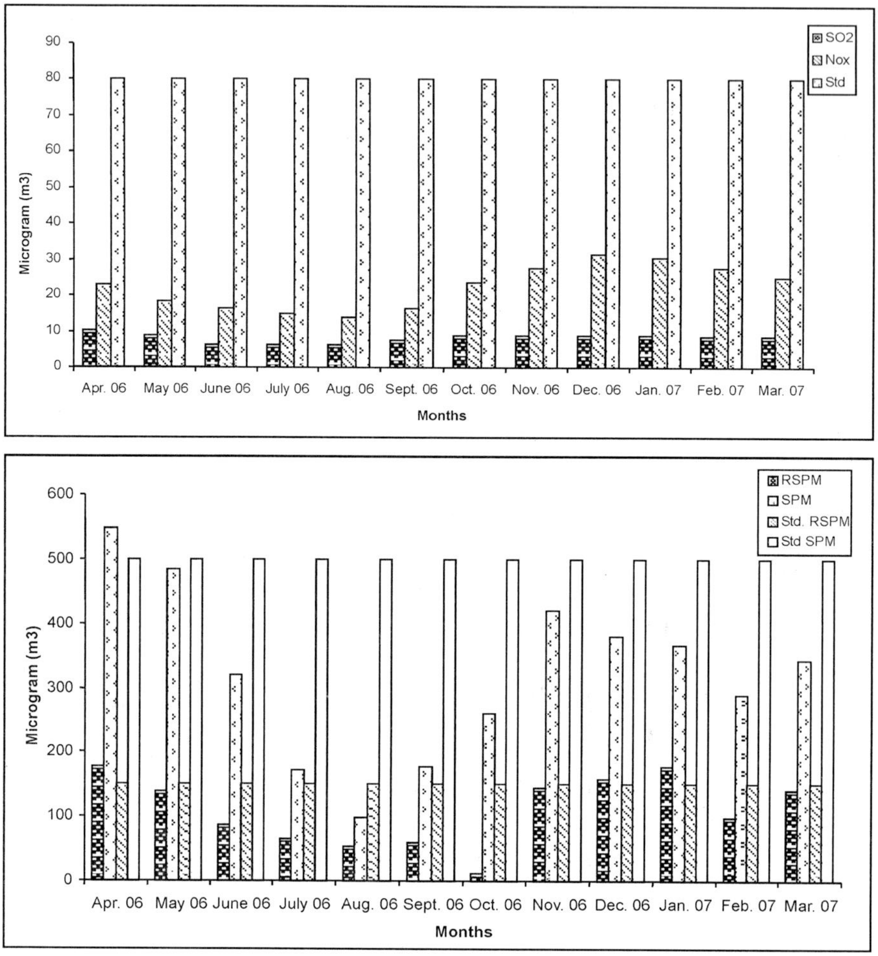

In the industrial area there is a concentration of various types of industries which is responsible for the high air pollution.

In Kota North also high polluted area is located. This is because of the railway station high concentration of air pollution and more vehicular traffic.

The various components of air pollution have been shown in Table 5.2.

Super Thermal Plant of Kota is the main cause of air pollution not only in adjoining areas of the plant but it also affects a large part of the city.

Other areas of high air pollution are located along main city roads such as Railway Station to Aerodrome Circle, and its extension Jhalawar Road, Aerodrome Circle to CAD Circle and DCM Road, Gumanpura-Chhawani Road, Bus Stand to Nayapura-Station road, etc. The main traffic points identified as high polluted points are:

1. Anta Ghar
2. Agarsen Chouraha
3. Nayaprua Chouraha
4. Bus Stand
5. Kota Junction
6. Chawani Chouraha
7. Aerodrome Circle
8. I.L. Chouraha
9. Keshavpura Chouraha
10. Mahaveer Nagar Circle
11. CAD Chouraha
12. Kethunipole
13. Kunhari Chouraha
14. Modi College Tiraha
15. Dadabari Tiraha

Table 5.2 *Air Pollution Components at Kota North*

Month and Year	SO_2	NO_2	*RSPM*	*SPM*
April, 06	137	399	137	399
May, 06	146	450	146	450
June, 06	72	202	72	202
July, 06	58	163	58	163
August, 06	25	70	25	70
September, 06	54	100	54	100
October, 06	98	198	98	198
November, 06	153	278	153	278
December, 06	166	277	166	277
January, 07	146	244	146	244
February, 07	90	180	90	180
March, 07	136	237	136	237

In Kota city number of automobiles are increasing day by day. At present the number of automobiles is 23,200 in the city. Another cause of air pollution is the National Highway No. 12, passing through the city. About 27,000 heavy vehicles pass daily through this highway causing air pollution. The outer areas of the city, the Military Area, Civil Lines and newly developed areas of Kota are having minimum air pollution.

Water Pollution

Water pollution simply means contamination of water due to any external material, or in other words, introduction of something to natural water which makes it unsuitable for human consumption. In general, 'water pollution may be defined as the adverse change in composition or condition of the water such that it becomes less suitable for which, it would be suitable in its natural state. The changes include physical changes, chemical changes and biological changes' (Manivasakam, 1992).

Kota is located at the bank of river Chambal, river Chambal is the only perennial river of Rajasthan. Thus, Kota city has an advantage of regular supply of water throughout the year. But at the same time river Chambal is now facing problem of pollution from the city itself. The whole waste water including municipal and domestic waste water is flowing into the river without any treatment. Therefore, the water of river Chambal now has been polluted. There are 22 *nalas* (drains) which used to pore dirty waste water in river Chambal. The following table indicates the drains and their location point at Chambal.

Table 5.3 *List of Drains (Nalas) at River Chambal*

S.No.	*Name of the Drain (Nala)*	*Location Point at River Chambal*
1.	Shivpura – Bheetriya Kund Nala	Near Bheetriya Kund
2.	Godawari Dham Nala	Near Godhwari Dham
3.	Amar Niwas Nala	In between Godawari Dham into river
4.	Brajdhar Nala	After Godawari Dham
5.	Near Borao-ki-Bagichi Nala	After Kota Barrage
6.	Kishorepura Badi Pulia Nala	Kishorpura
7.	Sajidehada Choti Pulia Nala	Kishorpura
8.	Nala of Bhario ji ka Chowk	Kishorpura
9.	Kabristan Nala	Kishorpura
10.	Adharshila Nala	Near Adharshila
11.	Ladpura Nala	Near Karbala
12.	Nayapura Nehru Colony Nala	Near Colony
13.	Civil Line Nala	Civil Line
14.	Nehru Nagar Nala, Kota Jn.	Near Nehru Nagar Basti
15.	Sanjay Nagar Nala, Kota Jn.	Near Sanjay Nagar
16.	Chandraghata Nala	Downward Chandraghata
17.	Patanpole Nala	Near Bhatt ji ka Ghat
18.	Khai Road Nala	Near Chambal Bridge
19.	Brajdhar Nala	Downward Patanpole
20.	Nayapura Nala	Downward Nayapura
21.	Khedli: Purohit ka Nala	Dowward Khedli
22.	Fatheghardi Rampura Nala	Bhatapada

All the above mentioned drains have become a major cause of water pollution of river Chambal.

The Rajasthan State Pollution Board has indicated an upward trend in respect of DO, BOD and COD at Akelgarh, Chambal Garden and Railway

Bridge. Similar trends have been observed in respect of TDS at various places. The BOD level of +15 mg/1 is much higher against prescribed limit of 2 and 3 mg/1. The DO level at Bhittriya Kund, Rampura, Chambal Garden, Nayapura Bridge and Railway Bridge is higher than permissible norms.

Apart from the above mentioned drains the water pollution by domestic waste and industrial waste water is also common. The water of Kishore Sagar is highly polluted. Although its water is not used for drinking purpose but because of its location in the middle of the city, it has become a cause of environmental degradation.

The water polluted areas of the city have been depicted in Figure 5.3. The highly water polluted areas include area along river Chambal (especially where city *nalas* pour their dirty water), areas of Kota North, areas of old Kota city. The minimum water pollution has been observed in ward no. 1, 4, 5, 7, 12, 15, 16, 22, 26, 32 to 37. The main problem is the increasing water

Figure 5.3 *Kota: Water Polluted Areas*

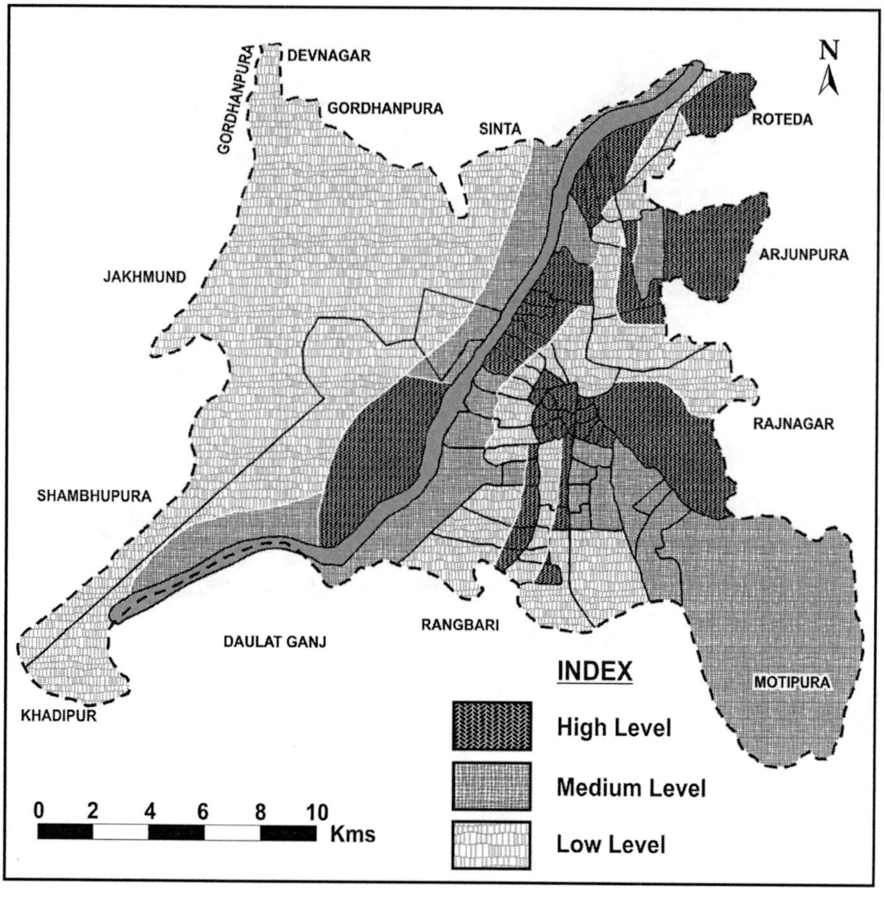

pollution in river Chambal, which is a great threat not only to the people of Kota city but also for the life of all those organisms living in Chambal river. A plan for the purification of Chambal river water has already been approved by the Central government. After completion of this project, the conditions will definitely improve.

Land Pollution

Land pollution is the result of misuse of land and can be defined as the changes in physical, chemical and biological conditions of the soil through man's intervention. Rapid urbanisation with the consequent increase in buildings resulted in the reduction of lands for the wastes to be disposed. Dumping of industrial and municipal wastes causes leaching by toxic materials which seep into the soil and affect the ground cause. It is also a cause of several diseases caused by bacteria. In fact, land and soil pollution is also a main cause of air and water pollution.

Figure 5.4 *Kota: Land Polluted Areas*

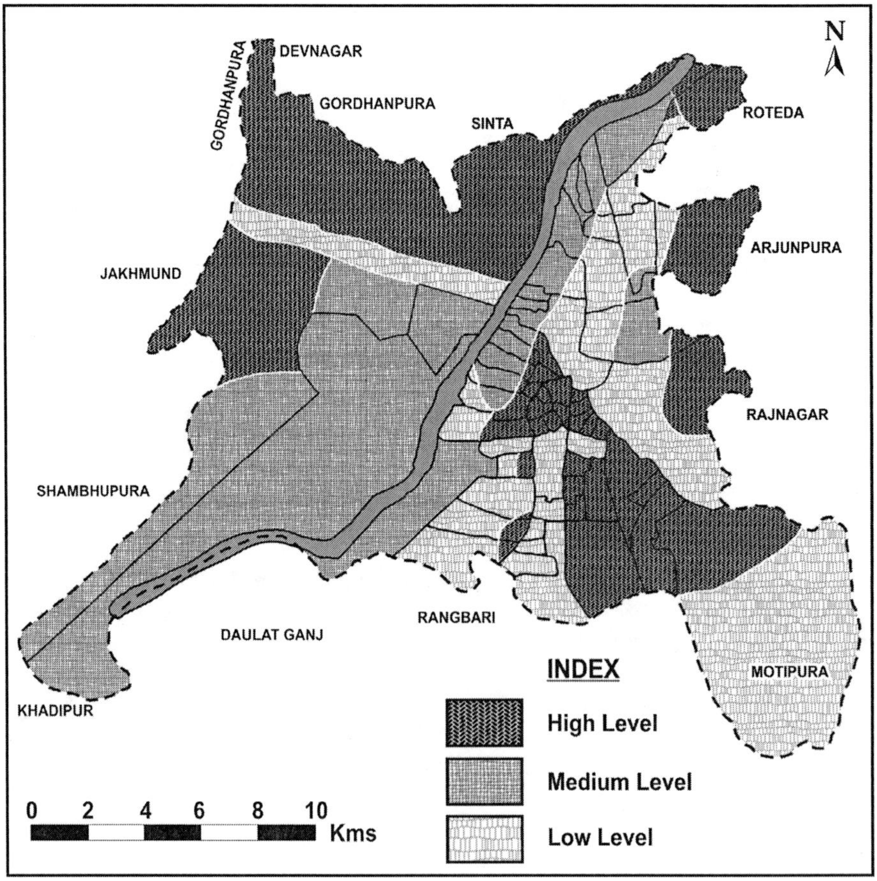

Like any other city of India, Kota is also having problem of land pollution mainly caused by domestic and municipal waste as well as by industrial wastes. The quantity of domestic and municipal waste is increasing day by day. At Kota urban wastes comprise both domestic and commercial wastes including dried sludge of sewage. Since data regarding land pollution are not available, the present analysis is based on detailed fieldwork. On the basis of fieldwork a map of land polluted areas of Kota city has been prepared (Figure 5.4).

The spatial pattern of land polluted areas shows that areas in which high land pollution has occurred are north-eastern part of the city, Kishorsagar to Chawani Chouraha extending in ward no. 41 to 54 and areas of ward no. 1, 13 and 17. The old Kota city is in the medium category of land pollution area. The areas along river Chambal and Sakatpura-Kunhari are also having medium land pollution. The minimum land pollution areas include military area of Kota Junction and colonies of new Kota. The main problem of land pollution of the city is unsatisfactory collection and disposal of domestic and municipal wastes.

Noise Pollution

Noise pollution is an environmental problem that has only recently received much attention. More and more noise is the creation of urbanisation and has now become a major environmental pollutant. The main source of noise pollution in cities is noise associated with transportation, automobiles, motor cycles, rail engines, buses, air planes, industries, high pitched musical instruments, etc.

Kota city is also suffering from noise pollution. Rajasthan State Pollution Control Board, Jaipur has calculated ambient noise levels and percentiles for various types of areas in Kota city that has been given in Table 5.4 and also shown in Figure 5.5.

Table 5.4 *Ambient Noise Levels and Percentiles for Various Area Types, Kota*

Area Type	*Noise Levels*		*Percentiles*									
	Day	*Night*	*Day*					*Night*				
			L_{10}	L_{30}	L_{50}	L_{70}	L_{90}	L_{10}	L_{30}	L_{50}	L_{70}	L_{90}
Residential	54	43	57	51	48	46	43	55	38	36	34	33
Commercial	68	59	71	65	62	59	52	60	53	50	46	41
Industrial	53	46	55	52	50	47	45	48	46	44	42	40
Sensitive	63	55	65	58	53	49	41	56	48	44	40	35

Figure 5.5 *Ambient Noise Level Percentiles Day and Night Time of Kota City*

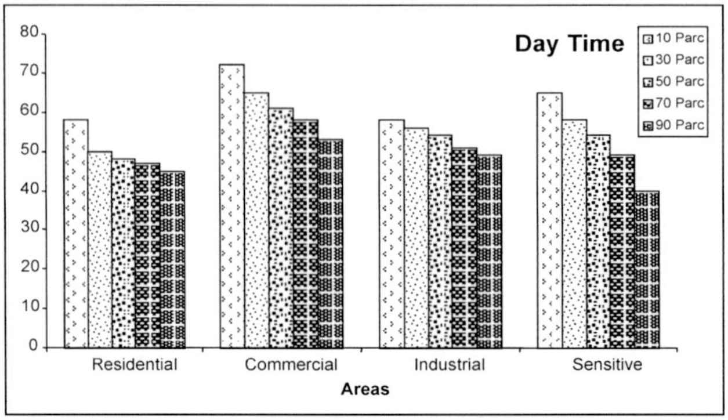

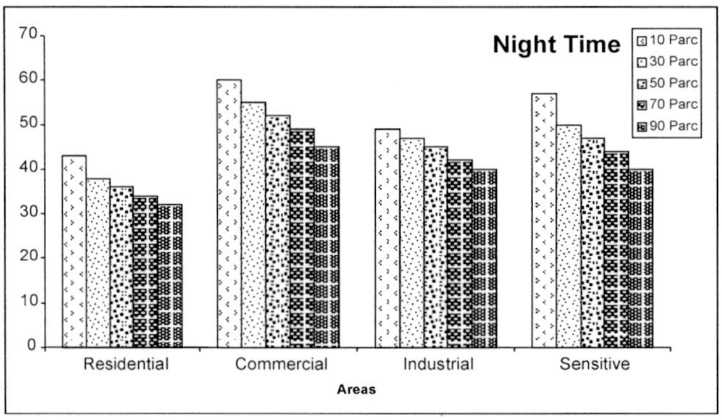

Figure 5.6 *Kota: Noise Polluted Areas*

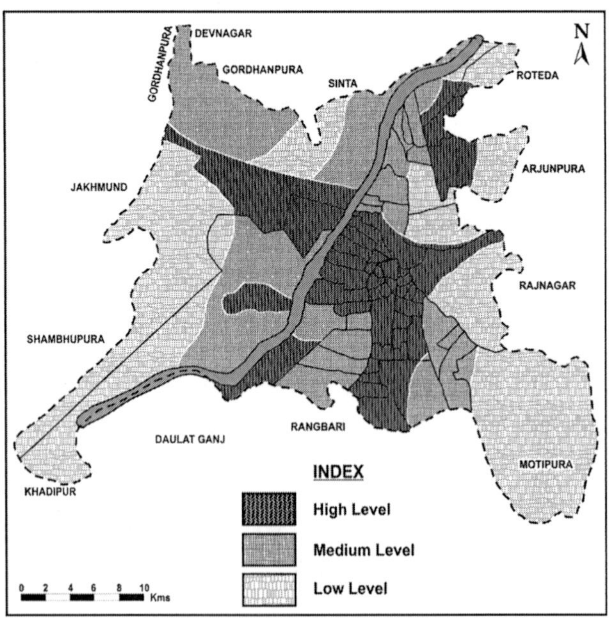

It becomes clear from the above table that the average level for residential areas during day time was 54 db, whereas, for night time it was 43 db, which is within the permissible limit. Except the sensitive area, all other areas of the city are within permissible limit. Based on fieldwork a map of noise polluted areas has been prepared (Figure 5.6). The map depicts that the highly noise polluted areas are located along National Highway No. 12 and 76. The old city of Kota and railway station areas are also in high noise pollution category. While the areas having minimum noise pollution include outer areas of Kota city and military area in North Kota. Rest of the city is under the medium noise polluted area.

Bharatpur

Bharatpur is a medium size town having 2,05,235 population in 2001. The population of the town is increasing at a fast rate, the growth rate during 1971–81 was 50.68 per cent and in 1981–91 it was 49.02 per cent. Although, growth rate of population has been reduced to 30.82 per cent during 1991–2001. But because of rapid growth of population as well as growth of city morphology and infrastructure, the problem of pollution has also increased, which should be taken into consideration for future planning and maintenance of the quality of life.

The spatial pattern of air, water, land and noise pollution in Bharatpur has been depicted in Figure 5.7, 5.8, 5.9 and 5.10 respectively. All these maps are based on fieldwork.

The spatial pattern of air pollution in Bharatpur city (Figure 5.7) indicates that areas having high air pollution are along the main roads, i.e. circular road within middle of the city, Jaipur road, Agra road and Mathura road. Jaipur-Agra road is National Highway No. 11 which passes through the southern part of the city and a main source of air pollution. The 14 areas have been identified as air polluted areas in Bharatpur city. These are as follows:

Saras Chouraha	(Ward No. 21)
Kumar Gate	(Ward No. 32)
Industrial Area	(Ward No. 42)
Hiradas Bus Stand	(Ward No. 1)
Surajpole Chouraha	(Ward No. 26)
Choburja	(Ward No. 14)
Lakshman Mandir Kamala Road	(Ward No. 30)
Bijlighar Chouraha	(Ward No. 20)

Raja Man Singh Circle	(Ward No. 19)
Bharatpur Junction	(Ward No. 43)
Red Cross Circle	(Ward No. 36)
New Hospital Tiraha	(Ward No. 2)
Sever Mentra Road	(Ward No. 1)
New Mandi Chouraha	(Ward No. 41)

Figure 5.7 *Bharatpur: Air Polluted Areas*

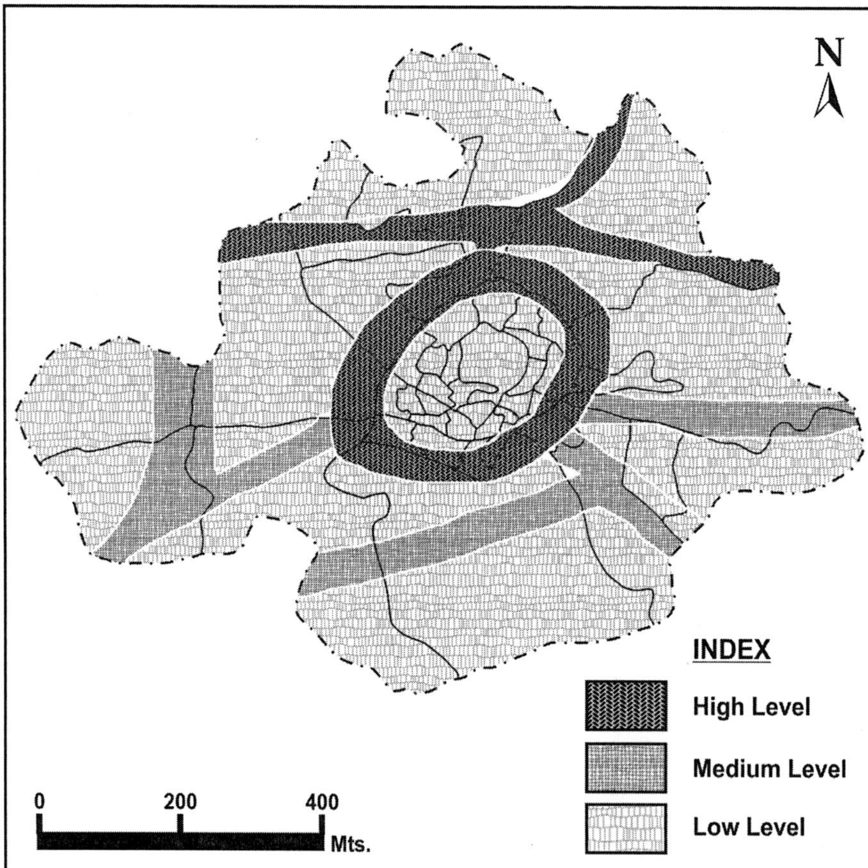

The water polluted areas of Bharatpur have been shown in Figure 5.8. The highly water polluted area is the Sujan Ganga Mahar and its adjoining area. This area is located in the heart of the city. The water polluted areas in ward no. 14 include Sarafa Bazar, Budh Ki Hat, Ganga Mandir Chouraha up to Choburja Chouraha. Several drains put their waste water in this canal. The waste of Keshav Nagar, Janana Hospital, Dutta Guest House, Jama Masjid, Agarwal Dharamshala, etc. has been poured in this canal which creates water pollution adjoining to this canal. The other areas having water pollution are Surjeet Nagar, Bajaj Nagar, Sardar Mohalla, Wasan Gate, etc. the western part of the city is also having problem of water pollution. Apart from above mentioned areas other areas of the city are not having much water pollution.

The land polluted areas of Bharatpur include (i) RIICO Area, (ii) Industrial Area, (iii) Behind M.S.J. College, (iv) Kumargate Grain Mandi, (v) Railway Colony and area adjoining to railway station, (vi) Marriage

Figure 5.8 *Bharatpur: Water Polluted Areas*

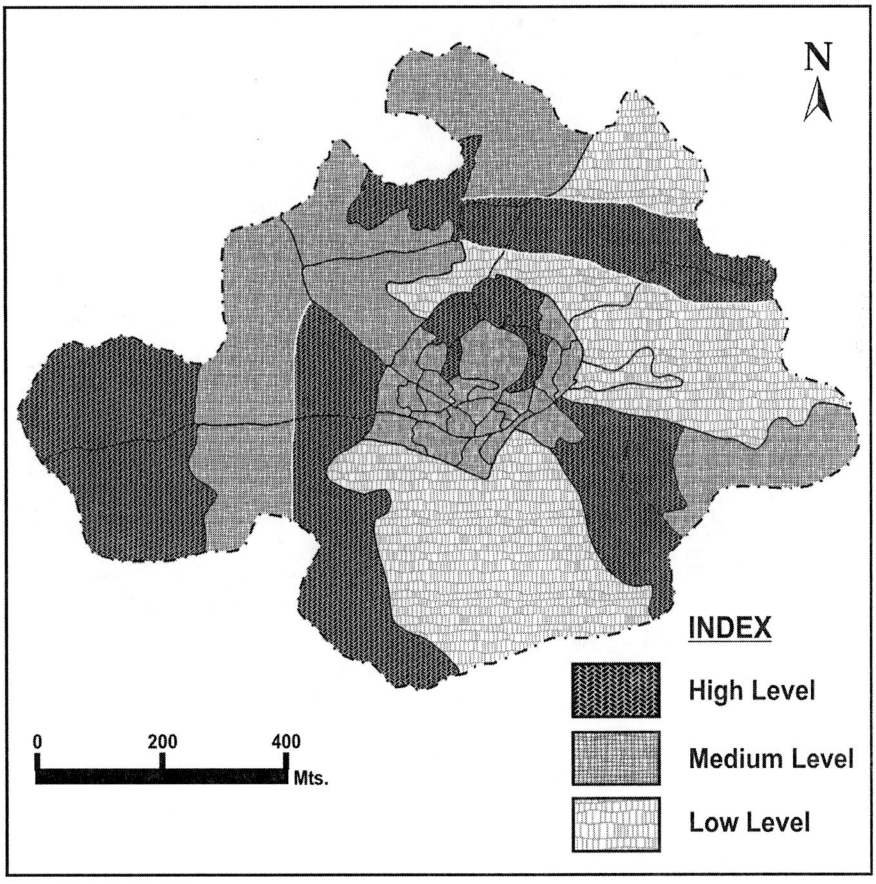

Figure 5.9 *Bharatpur: Land Polluted Areas*

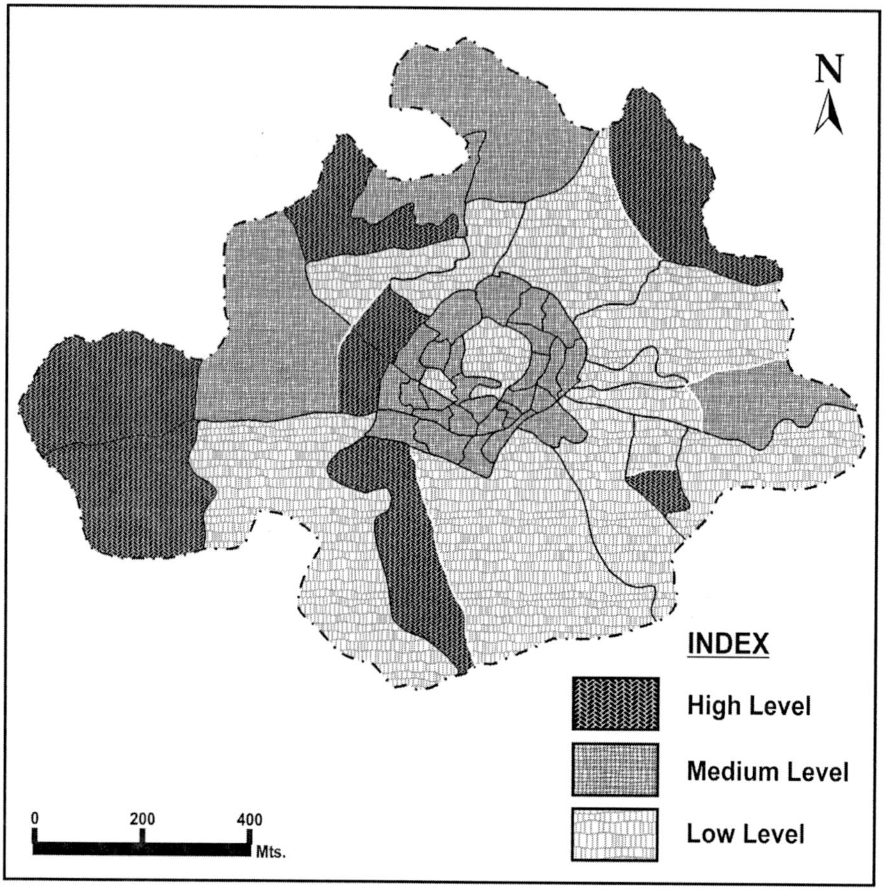

Home Area, (vii) Company Bag Park, (viii) Ekta Vihar, (ix) Area adjoining to Sujan Ganga Nagar, (x) Kamalpura, (xi) Jatav Basti, (xii) Harijan Basti, (xiii) Dhimar Mohalla, (xiv) Kali Bagichi, and (xv) Sanjay Nagar (Figure 5.9). The main problem of land pollution is unsatisfactory arrangement of collection and disposal of domestic and municipal waste.

The various categories of area of noise pollution, i.e. high, medium and low have been depicted in Figure 5.10. In Bharatpur City 17 areas have been identified as noise polluted areas. These are Hiradas Bus Stand, Simco Area, Railway Station, Choburja Chouraha, Laxman Mandir Area, Saras Chouraha, Bijlighar Chouraha, Kumhar Gate Area, Red Circle, Ganga Mandir, Maduvan Capital Area, Binasagan Gate, Anah Top Circle, Atal Band, New Grainmandi Chouraha and Gordhan Gate.

The problem of noise pollution is not so acute, but for future planning its control is necessary.

Figure 5.10 *Bharatpur: Noise Polluted Areas*

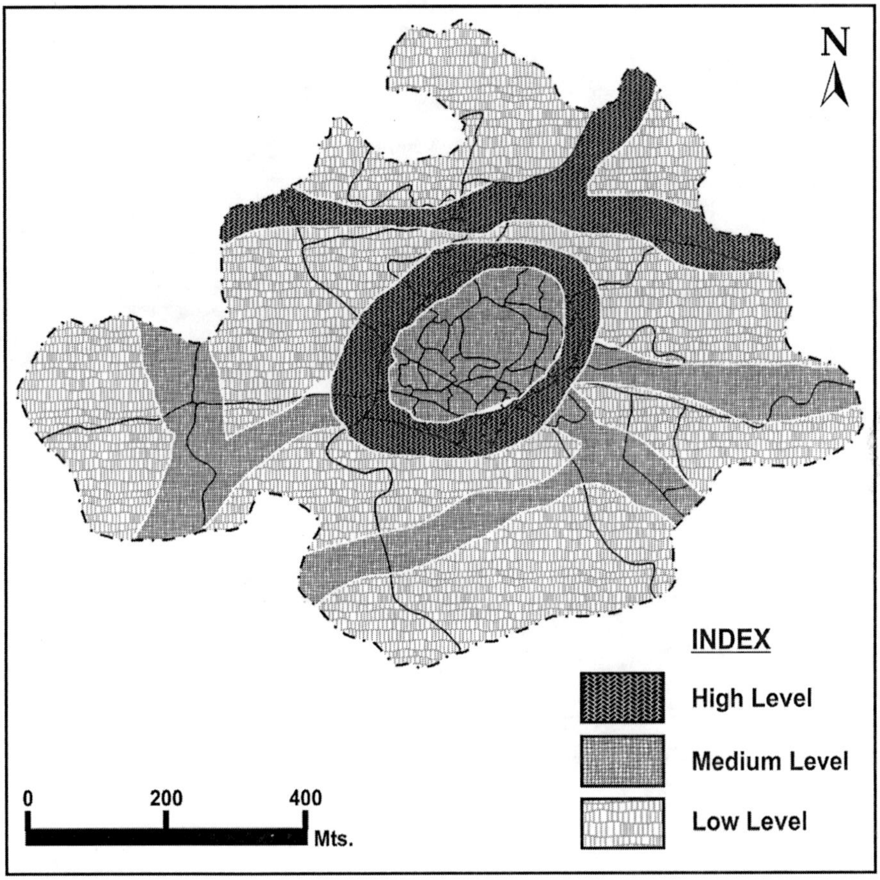

Sriganganagar

Sriganganagar is a newly developed town of Rajasthan. The town was developed after the advent of Gang Canal in 1927. It is a planned town and having population of 2,22,833 in 2001. It is a fast growing town and an important market centre of the northern Rajasthan. In spite of planned growth, several areas of the town are under the grip of pollution. The spatial pattern of air, water, land and noise pollution has been depicted in Figures 5.11, 5.12, 5.13 and 5.14 respectively. All these figures are based on fieldwork, because information regarding pollution is not available.

The spatial pattern of air pollution in Sriganganagar as shown in Figure 5.11 indicates that there are two areas having high degree of air pollution. The first is Sugar Mill area and second is the J.C.T. Mill area of Purani Abadi. Sugar Mill is responsible for air pollution in east central part of the town, i.e.

Figure 5.11 *Sriganganagar: Air Polluted Areas*

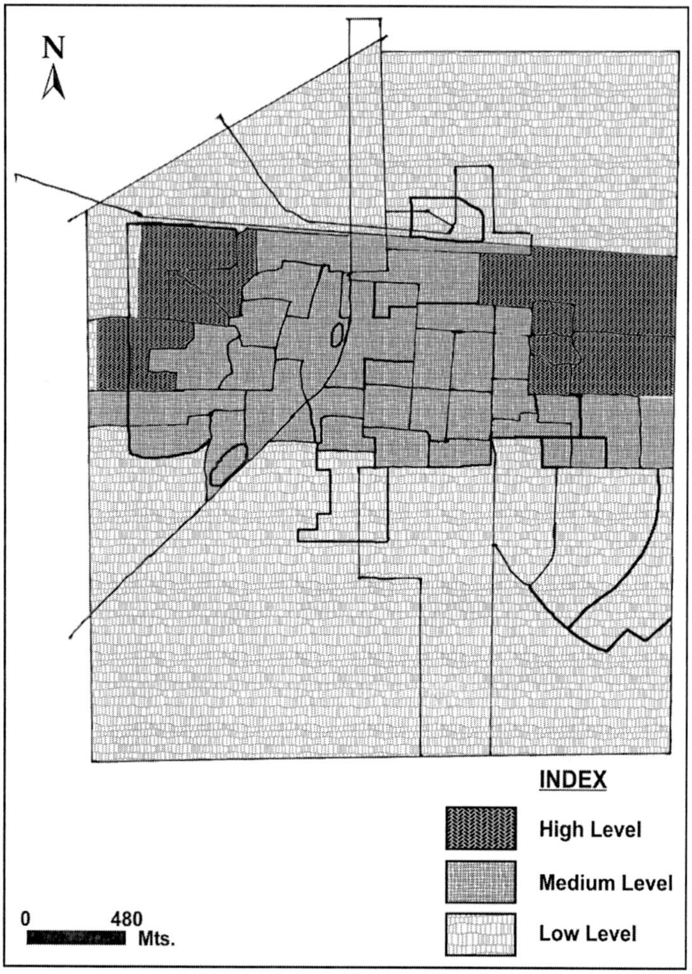

in ward no. 3, 9, 10, 11 and 12. The eastward winds often extend the pollution area westward. The J.C.T. Mill is a cause of air pollution in Purani Abadi in ward no. 35, 36, 37 and 30. In between these two high areas the central part of the city is in medium air polluted area. The whole of northern and southern area of the town is having lesser air pollution.

The high degree water polluted areas include Sugar Mill, Near J.C.T. Mill, Guru Nanak Basti and Sadbhawana Nagar. In Sugar Mill area there is a huge depression in which water from Sugar Mill as well as from adjoining areas is filled up and becomes a highly water polluted area. Also near J.C.T. Mill there are several small depressions filled with dirty water and creates water pollution. In Sadbhawana Nagar located in ward no. 24 in southern part of the town waste water of town is collected but since there is no provision of its treatment,

Figure 5.12 *Sriganganagar: Water Polluted Areas*

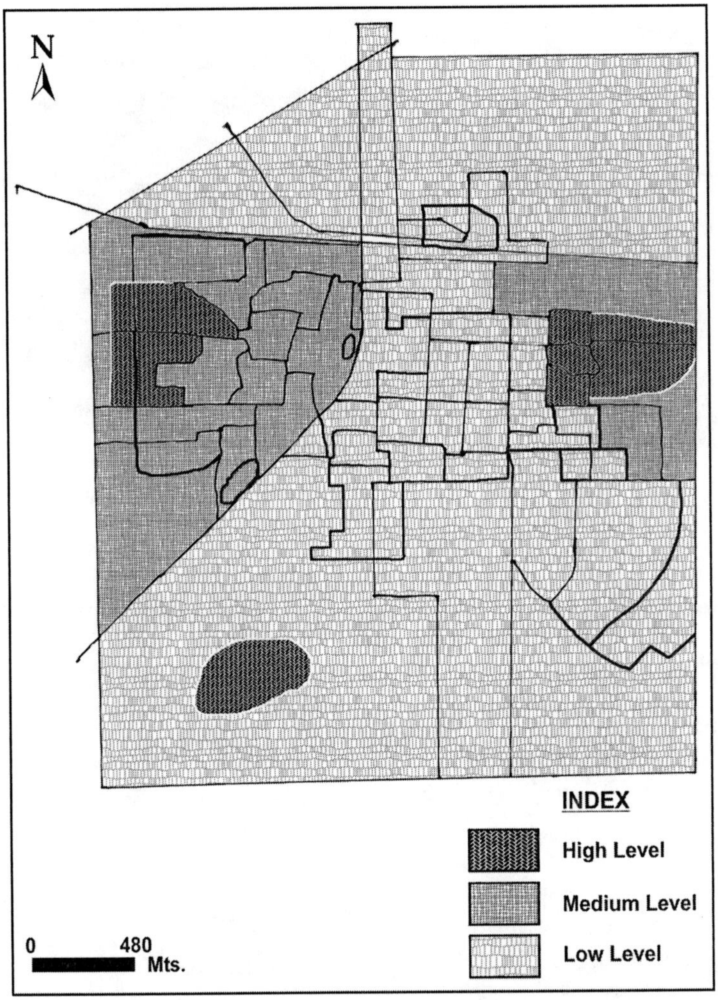

it becomes a main cause of water pollution. The whole of the Purani Abadi area is having water pollution problem. Another area is adjoining to high polluted area of Sugar Mill. Rest of the city, which includes a large part of the city is having lesser problem of water pollution (Figure 5.12).

The land polluted areas of Sriganganagar has been shown in Figure 5.13. A large area of city is suffering from land pollution. Whole of Purani Abadi including areas of ward no. 26 to 39 and half of ward no. 40 are highly land polluted because of the dumping of domestic and municipal waste at several places. Another large area of land pollution is the north-east and east-central part of the city.

Figure 5.13 *Sriganganagar: Land Polluted Areas*

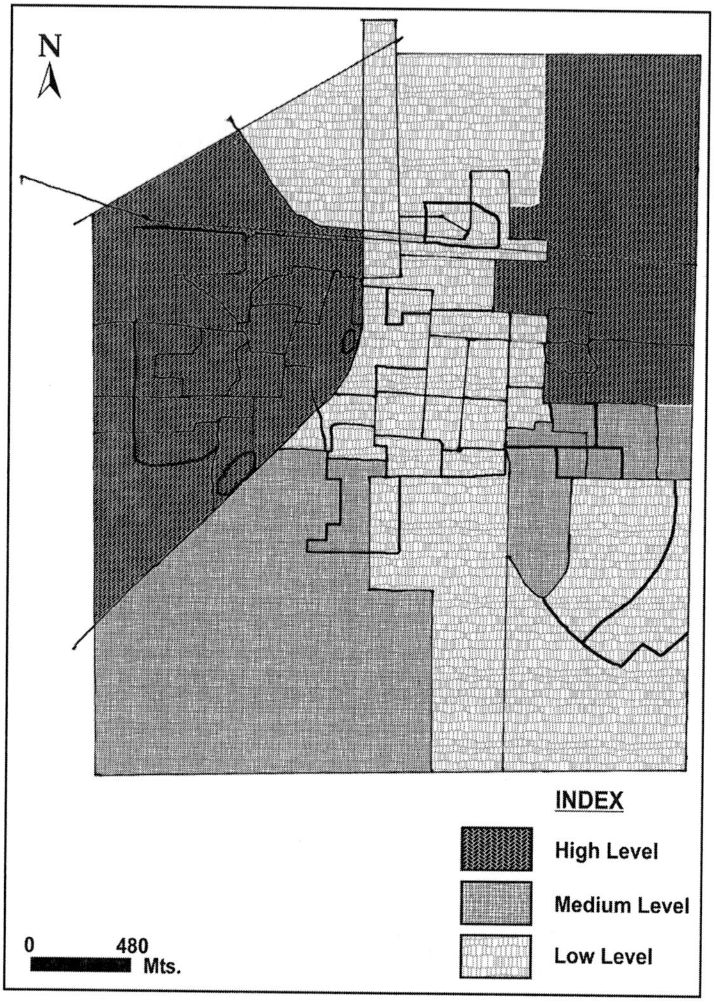

The medium land polluted areas are located in ward no. 15, 17 and 24. While most of the Block areas and newly developed colonies located in south-eastern part are having lesser land pollution.

The spatial pattern of noise pollution at Sriganganagar is having a different pattern. The high noise polluted area starts from railway station extending in Block area and finally extending along Hanumangarh road. The medium noise polluted area is Sugar Mill area and area of Purani Abadi from railway crossing of Hindumal Kot to Bus Stand. All other areas of north, west and southern part of the city are not having much noise pollution (Figure 5.14).

Figure 5.14 *Sriganganagar: Noise Polluted Areas*

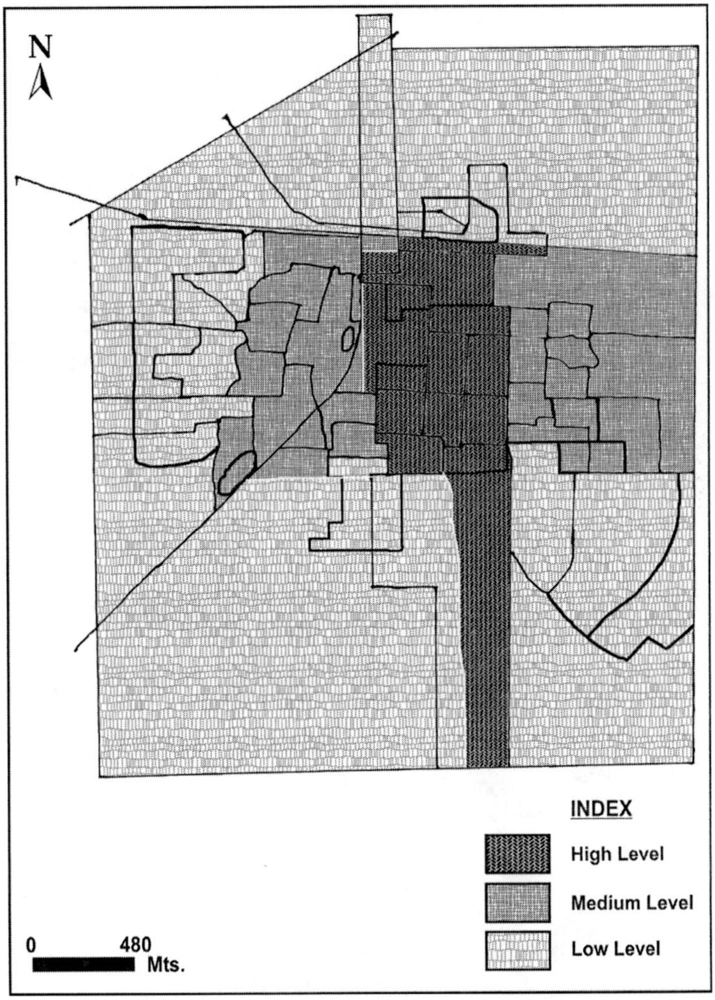

Tonk

Tonk is a small town of Rajasthan having population of 1,35,689 in 2001. Located at the bank of river Banas, Tonk is a traditional type of town and its growth rate is also slow in comparison to other towns of the State. The spatial pattern of air, water, land and noise pollution in Tonk has been depicted in Figure 5.15, 5.16, 5.17 and 5.18 respectively. The depiction is based on field survey.

The spatial pattern of air pollution as shown in Figure 5.15 indicates that two specific areas are highly polluted. These are along National Highway No. 12 and a central belt extending from Clock Tower to Modern

Figure 5.15 *Tonk: Air Polluted Areas*

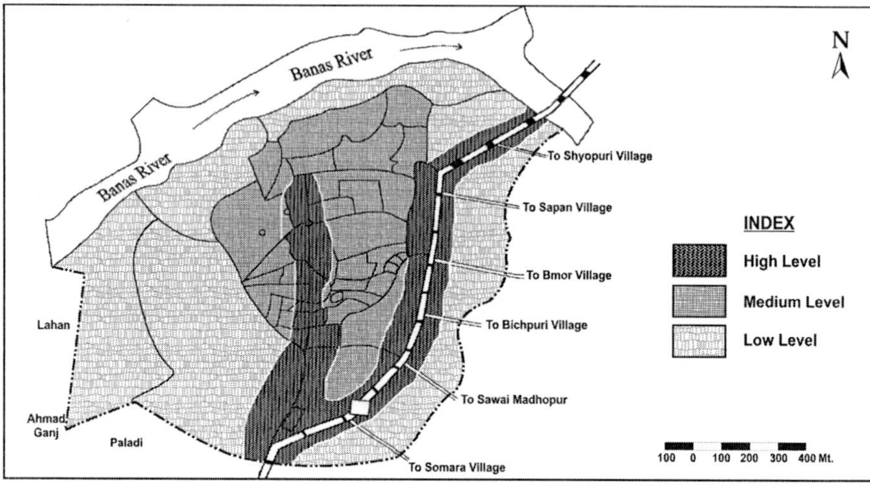

factory and joins NH–12. The reason for high pollution along NH–12 is the heavy rush of trucks, buses and other automobiles passing through the highway. While other polluted areas are roadways depot, Bada Kua, Panch Batti area, Ghantaghar area and Bus Stand road. On the other hand medium air polluted areas are adjacent to the high areas. The outer sides of the town is having low level air pollution. A few industries located in Tonk are also a cause of air pollution. A slaughter house located at Kamela has become a main cause of air pollution in this area. Similarly fish market of Tonk is also highly polluted.

Water pollution is a serious problem of Tonk. The water polluted areas of Tonk have been depicted in Figure 5.16. It becomes clear from the map that tanks of the city are responsible for water pollution. The areas around these tanks are highly water polluted areas. There are eight tanks in the city. These are Mahadevwali Talab, Sankhali Talab, Idghah Talab, Radiwas Talab, Dhanna Talai, Annapurana Talab, Taliyan Talab and Chaturbuj Talab. The whole Purana Tonk is also suffering from water pollution problem. While the south-western, southern and eastern part of the city is having low level of water pollution. Rajasthan State Pollution Control Board (RPCB) has analysed the water quality and reached to the conclusion that the trends of DO and BOD are by and large static, whereas COD tends to show an upward trend with a rise of 8 mg/1 of COD. The trends in respect of TDS shows a downward trend of TDS with a fall of +400 mg/1.

The main cause of land pollution in Tonk is domestic and municipal waste. Since the arrangements of collection and disposal of wastes are not proper they create problem of land pollution. This problem is high in whole

Figure 5.16 *Tonk: Water Polluted Areas*

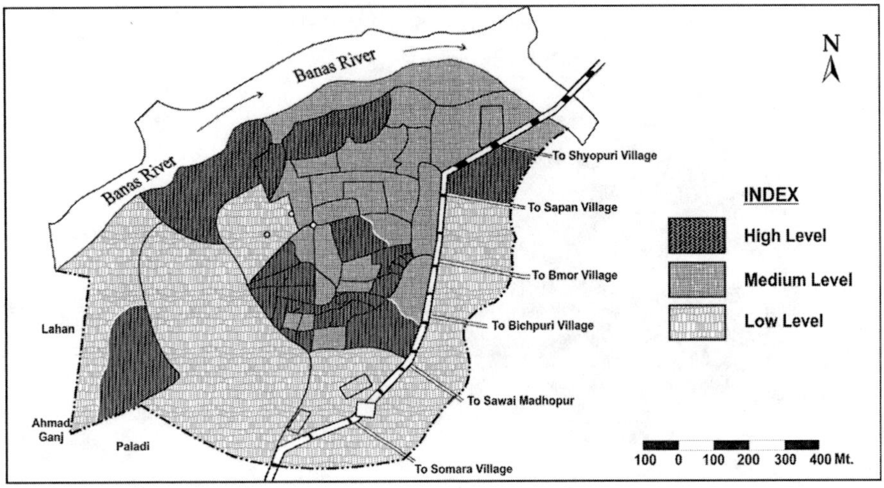

of the old city, while medium along river Banas and lower in eastern and mid-west of the town (Figure 5.17).

The areas facing problem of noise pollution have been shown in Figure 5.18. The main area is along National Highway No. 12 and area extending from Ghantaghar in the north to NH–12 in the south. The areas having more noise pollution are roadways depot area, Bada Kua area, Panch Batti area, Ghantaghar area, Qafala Bazar, Naushemian ka Pul and Subhash Bazar. The low noise polluted areas are extending from south-western to entire eastern part of the city. In between these two medium noise polluted areas are located.

Figure 5.17 *Tonk: Land Polluted Areas*

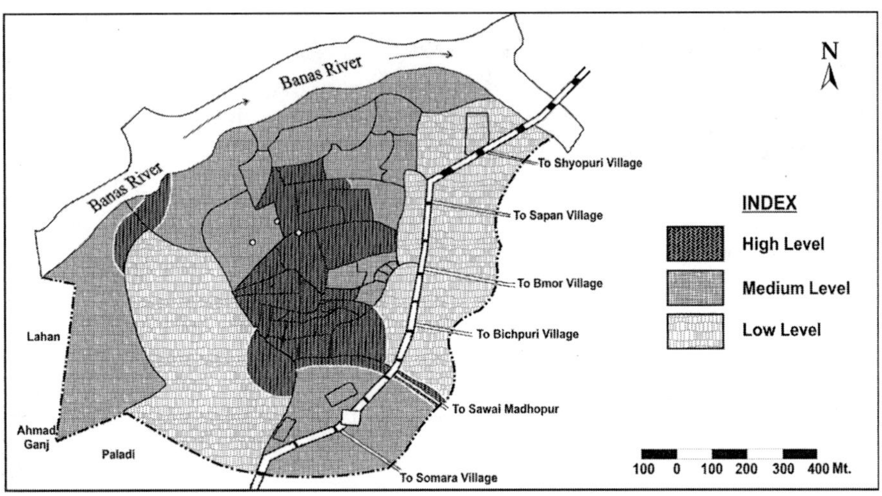

Figure 5.18 *Tonk: Noise Polluted Areas*

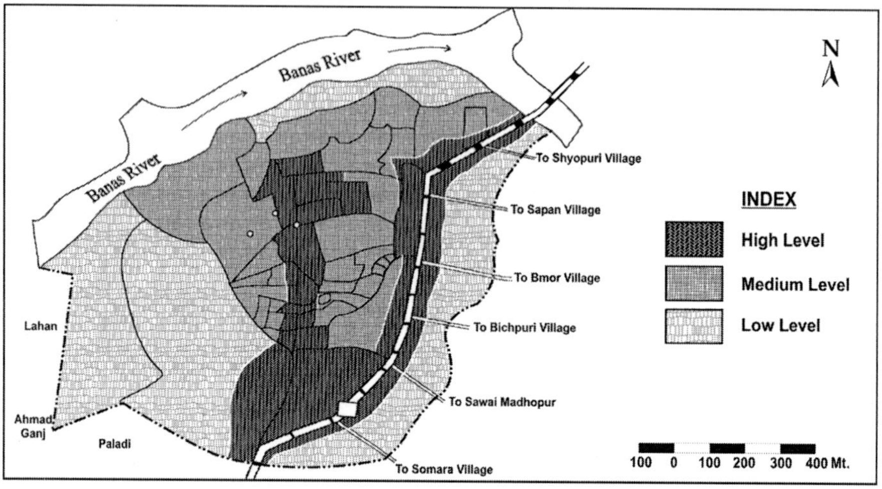

Gangapur City

Although its name is Gangapur City but it is a small town having population of 1,05,336 in 2001. The town is located along Delhi-Kota-Bombay railway line and is a small commercial mart of Sawai Madhopur district. Information regarding pollution of this town is not available. Therefore, the present analysis is based on field survey. Gangapur City has been selected for pollution study because it will explain the nature of pollution in towns just having above one lakh population. The spatial pattern of air, water, land and noise pollution in Gangapur City has been depicted in Figure 5.19, 5.20, 5.21 and 5.22 respectively.

The areas having high air pollution are industrial area of the town in which about ten factories are located. The other areas are where air pollution is due to automobiles, such areas are Truck Union area, Favarra Chowk, Bus Stand, Railway Station, Gas Godown area and Udai Mod. The areas having medium air pollution are located in south-western, southern, northern and also in central part of the town (Figure 5.19).

The problem of water pollution in Gangapur City is quite moderate (Figure 5.20). Only Naya Bazar, Ram Bag to Kranti Nagar, Kailash Talkies to Dasherra Maidan and Chuli Gate to Pani Naria colony are having problem of water pollution. Otherwise most of the town is not having any serious problem of water pollution.

The land pollution is also limited in the city. The high land polluted area is located in ward no. 39, while medium land polluted area includes ward no. 4 and 36, rest of the Gangapur City is not having much land pollution (Figure 5.21).

Figure 5.19 *Gangapur City: Air Polluted Areas*

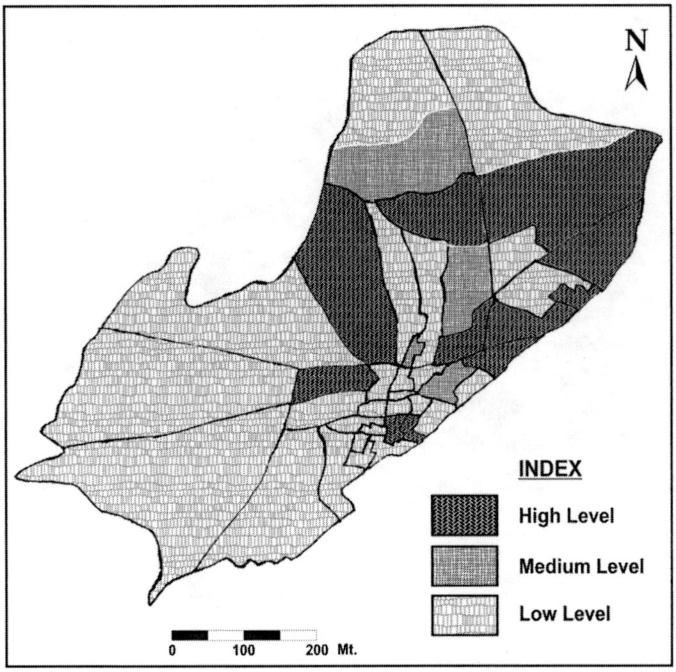

Figure 5.20 *Gangapur City: Water Polluted Areas*

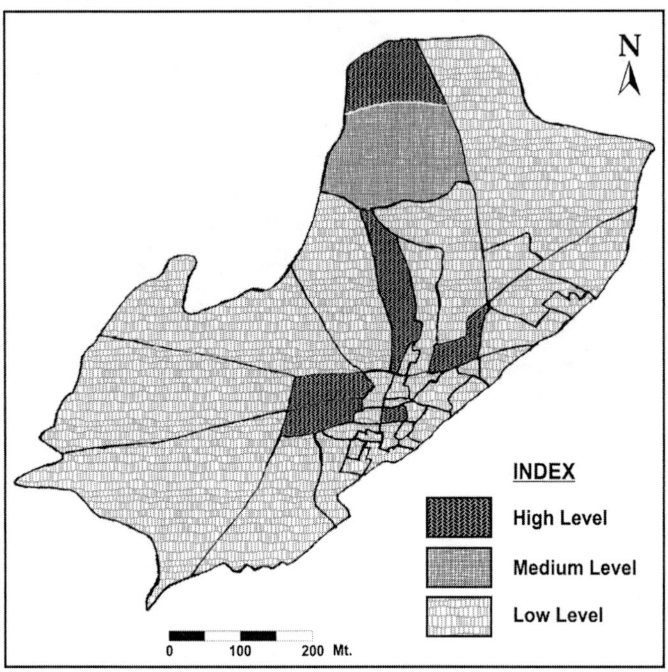

Figure 5.21 *Gangapur City: Land Polluted Areas*

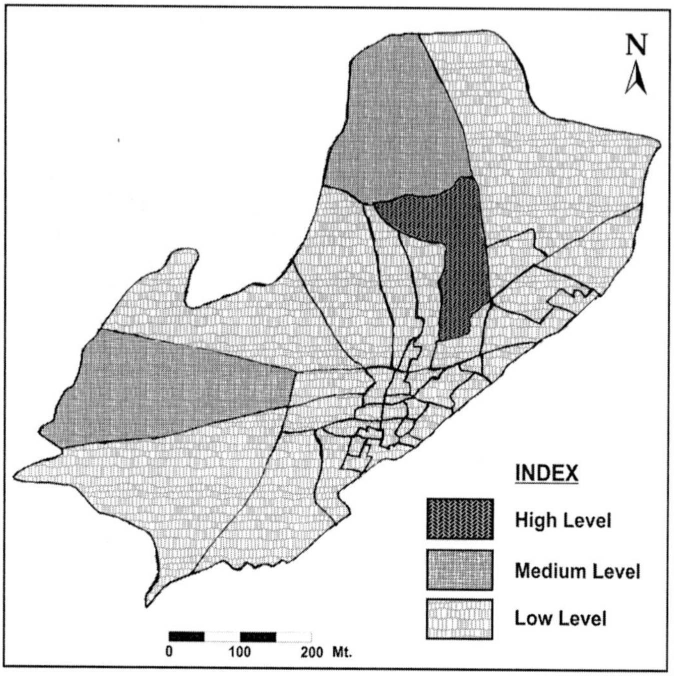

Figure 5.22 *Gangapur City: Noise Polluted Areas*

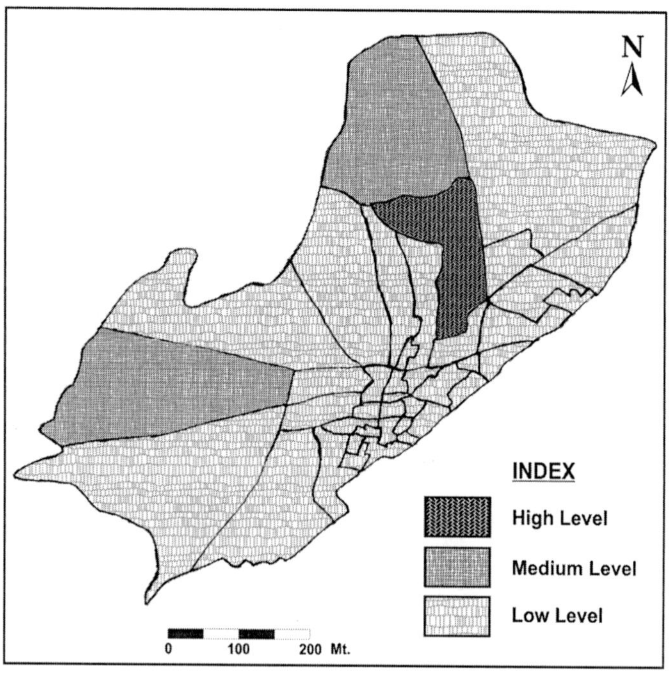

Like other polluted areas, noise polluted areas of Gangapur City are also limited. The high noise polluted areas include Truck Union, Fabbara Chowk, Railway Station and Bus Stand. Rest of the city is not having much noise pollution (Figure 5.22).

Since Gangapur City is a small town, problem of pollution is limited. But with the growth of town it may increase, therefore, attention should be given for its management.

References

Eyles, J. (1982), 'Health and Medicine in Urban Society: The Social Construction and Fetishism of Health, Contemporary Perspective on Health and Healthcare', Occasional Paper No. 20, Department of Geography, Queen Mary College, University of London.

Giggs, J.A. (1979), 'Human Health Problems in Urban Areas', in D.T. Herbert (ed.), *Social Problems and the City: Geographical Perspectives*, Oxford University Press, Oxford.

Manivasakam, N. (1992), *Environmental Pollution*, National Book Trust, New Delhi, p. 34.

Strahler, A.N. and A.H. Strahler (1977), *Geography of Man's Environment*, John Wiley & Sons, New York, p. 54.

Surana, B.K. and A. Bhargava (1997a), *Ambient Noise Levels in Rajasthan*, Rajasthan State Pollution Control Board, Jaipur.

— (1997b), *Assessment of Water Quality in Rajasthan*, Rajasthan State Pollution Control Board, Jaipur.

— (1997c), *Assessment of Ambient Air Quality in Rajasthan*, Rajasthan State Pollution Control Board, Jaipur.

6 Spatial Pattern of Quality of Life (QOL) in Urban Centres of Rajasthan

The concept of QOL is related with the welfare theme in geography or with *welfare geography*. 'The objective of welfare geography is the evaluation of the social desirability of alternative geographical status' (Henderson and Quandt, 1958: 201). While Nath (1973: 711) has defined, 'welfare geography is that part of geography where we study the possible effects of various geographical policies on the welfare of society'. The term *geographical state* or *situation* refers to the spatial arrangement of human existence, which may relate to spatial allocation of resources, income or any other aspect of human well-being, including quality of life. Therefore, geographical study of QOL not only explains the spatial pattern of hierarchy but also provides a base for planning.

The combination and fulfillment of needs and wants determined the QOL and motivation for human actions, from buying a food item to planning a city. The most 'basic' human needs are those related to physical survival, i.e. food, shelter, clothing, safe drinking water, etc. With the social, economic and technological growth human needs are not only diversified in the form of health, education, recreation, etc. but also upgraded day by day. Therefore, the standard of QOL varies with time and space. There are societies which are still struggling for survival, while advanced societies are more concerned with self-actualisation or personal fulfillment. The spatial variations in QOL are not only seen at national and regional level but also at local level. Such variations are more in cities, where population concentration is more. Therefore a criterion for the identification and measurement of the QOL in urban areas is to be determined before actual analysis of the spatial pattern of QOL in urban areas.

Criteria for the Determination of Urban QOL

The quality of life is a relative term, varies from nation to nation, society to society as well as from person to person, therefore, it is very difficult to develop a uniform criterion. David M. Smith (1977: 31) has rightly observed, 'the idea of trying to measure an abstraction like the quality of life is anathema to many people, even perhaps an assault on the individual'. But attempts have been made and still going on in geographical studies.

The studies done by Wilson (1969), Drewnowski (1970), Detwylor (1971), Cox (1972), Dickinson, Gray and Smith (1972), Howe (1972), Liu (1973), Ontell (1973), Berry (1973), Bederman (1974), Wilson (1974), Eyles (1974), Knox (1975), Terleckyj (1975), Koelle (1974), Duncan (1976), etc. have laid down the foundation of the geographical studies of QOL. They have used several variables for the determination of the levels of QOL in various parts of the world. A brief review of their work will help in determination of a general criteria.

Environmental Protection Agency (1973) has applied a procedure known as 'Delphi' to identify group value judgements with respect to what determines the quality of life. The criterion adopted by Dalkey and Rourke is based on the following variables:

1. Demographic process
2. Public participation
3. Health
4. Choices in life
5. Housing
6. Economic security
7. Education
8. Land use planning
9. Essential living cost
10. Economic opportunity

Berlin Centre for Future Research has developed a criterion for the measurement of '*Quality of Life*' under the direction of Koelle (1974). The purpose of this study was to arrive at an operational definition of the quality of life by generating a list of weighted goals. These are as follows:

	Primary Goals	*Weights*
1.	Improvement of the material quality of life	185
2.	Improvement of the physical quality of life	269
3.	Improvement of the mental quality of life	288
4.	Improvement of the spiritual quality of life	258
	Total	1,000

Each primary goal has been divided into four interim goals. On the basis of above methodology studies have been done in Germany and some European countries. But the main difficulty with this kind of analysis is that the preferences expressed may be un-representative.

Organisation for Economic Co-operation and Development (OECD, 1973) has developed social indicators for human well-being and QOL. The indicators selected by them are:

1. Health
2. Individual development through learning
3. Employment and quality of working life
4. Time and leisure
5. Command over goods and services
6. Physical environment
7. Personal safety and administration of justice
8. Social opportunity and participation

Drewnowski's approach is based on state of well-being which determined the QOL. His well-being index is as follows:

1. Somatic status (physical development level)
   - (i) Nutritional status
   - (ii) Health status
   - (iii) Life expectancy
   - (iv) Physical fitness
2. Educational status (mental development)
   - (i) Literacy
   - (ii) Educational attainment
   - (iii) Congruence of education with manpower requirements
   - (iv) Employment
3. Social status (social integration and participation)
   - (i) Integration
   - (ii) Participation

The above three major components are reflecting the physical, mental and social status of the people. He has also developed a level of living index having nine indicators. Both these indexes determined the quality of life.

Criteria of Social Well-Being in USA

In United States, social well-being is determined on the basis of following indicators (Smith, 1973):

1. Income, wealth and employment
2. The living environment
3. Health
4. Education
5. Social order
6. Social belonging
7. Recreation and leisure

These elements not only determined the social well-being but also the essence of life quality and this can help in identification of QOL at two different spatial scales – national and within the metropolis. While Bederman (1974) has used five variables to generate a quality of life index for Atlanta. These variables are health factor, public order factor, housing of quality factor, socio-economic factor and density factor.

Determination of Urban QOL in Indian Conditions

It becomes clear from above mentioned studies, that variables used by geographers for determination of the QOL are varied in nature and related with physical environment, income, health and nutrition, employment, leisure, educational status, security, social status, etc. All these studies have been done in developed countries, where conditions of urbanisation and standard of QOL is much higher in comparison to developing countries like India. In India, urbanisation has taken place at a fast rate but living conditions in cities are still not proper in terms of QOL and these conditions are entirely different than urban conditions of the developed countries. Therefore, the criterion used for the determination of QOL cannot be used in original form. This is also because of the fact that in our country detailed data about individuals/families are not available.

A framework for the determination and measurement of the quality of life has been designed for Indian cities (excluding metros) in general and cities of Rajasthan in particular. This framework is based on earlier studies but modified according to Indian conditions. The spatial pattern of the urban QOL can be determined on the basis of following variables:

(a) Housing conditions
(b) Nature of roads
(c) Drinking water facilities
(d) Sewerage/drainage system
(e) Waste disposal facilities

Along with above variables, income, literacy, health, density of population and leisure factors are also taken into consideration. The importance of these variables is as follows:

Housing is important as a source of shelter, comfort and social status; as 'home' it performs basic protective and symbolic function. Nature of house, its built, size and ventilation facilities have direct impact on QOL. Similarly neighbourhood quality covers such questions as whether the immediate environment outside the home is attractive or ugly, safe or threatening.

Nature of roads, i.e. cemented, dabur, paved or graveled indicate the overall development of a particular part of the city.

Drinking water facilities, i.e. availability of drinking water and its quality have had direct impact on human health.

Sewerage or internal drainage system is a primary requisite for an urban area. An absence of proper sewerage system adversely effect the QOL.

Waste disposal is the main problem of cities. The dumping of domestic and commercial waste in residential areas is also a main cause of environmental pollution and diseases, thus, having adverse effect on quality of life.

All the above factors along with *employment, income, literacy, health* and *leisure* determined the quality of life. Although, these factors are considered independently but in totality they not only direct but determined the spatial pattern of the urban quality of life.

Methodology

For the determination of the spatial pattern of QOL in cities having above one lakh population in Rajasthan, the variables selected are as follows:

1. Housing conditions
2. Street conditions
3. Drinking water facilities
4. Sewerage for internal drainage facilities
5. Waste disposal facilities

The nature and importance has already been discussed above. A detailed fieldwork has been conducted in sample cities regarding the existing pattern. The identification of areas having high, medium and low QOL has been determined by assigning weights for each category. The determination is like this.

Above 70 per cent	–	High
50 to 70 per cent	–	Medium
and below 50 per cent	–	Low

For each area/ward and for each related variable above mentioned category of QOL has been determined. Their spatial pattern has been depicted on city maps, which shows QOL based on that particular variable, i.e. housing conditions, drinking water facilities, etc.

According to 2001 census, Rajasthan is having 20 cities having above one lakh population, ranges from 1,00,476 of Jhunjhunu to 23,24,319 of Jaipur. Since it is not possible to study all the 20 cities, sampling has been done. In sampling, any specific technique has not been followed, but on the basis of total population selection has been done in such a way that they represent a specific class of the city. Thus, the samples selected are as follows:

1. Cities having above five lakh population – Kota (7,04,731) and Bikaner (5,29,007)
2. Cities having population between two to five lakh – Sriganganagar (2,22,833) and Bharatpur (2,05,104)
3. Cities having below two lakh population – Tonk (1,35,663)

The above samples represent not only the population band but are also located in different geo-economic regions of the State.

Spatial Pattern of QOL in Kota

Kota (76°10'E; 25°11'N) is located at the bank of river Chambal and is one of the most important cities of the State of Rajasthan. The importance of

Kota lies in the fact that (i) it is an administrative centre, i.e. district as well as divisional headquarter, (ii) an important market centre not only of Hadoti region but of Rajasthan, (iii) an industrial centre of the State, (iv) well connected with rail and road transportation with other parts of the country, (v) an important educational centre of all India status for the preparation of various examinations, and (vi) ranks third in terms of population among cities of Rajasthan after Jaipur and Jodhpur, etc. Historically also the city have had its importance. Kota is among the fast growing cities of Rajasthan and according to 2001 census its total population was 7,04,731. The highest growth rate of population was recorded during 1951–61 decade, when it was 84.84 per cent followed by 76.98 per cent in 1961–71 decade. During 1971–81 it was 68.20 per cent and in 1981–91 it was 50.00 per cent. But in 1991–2001 the decennial growth rate was 29.21 per cent. Kota city has recorded a rapid growth in its morphology and during last 20 years the city has recorded a fast growth in sectors like residential, commercial, transportation, power sector, industries, etc. All these factors have had their impact on environment and quality of life.

A detailed survey has been conducted for the study of QOL based on housing conditions, waste disposal facilities, drinking water facilities, internal drainage facilities and street conditions. For the purpose of fieldwork and data collection Kota city has been divided into three sectors, i.e. Kota North, Kota South or New Kota and Old Kota city. In all 60 wards of the city a detailed survey has been conducted. On the basis of survey and collected data, the analysis of QOL in Kota is as follows:

QOL Based on Housing Conditions

The best measure of urban QOL is the housing conditions. In each ward of Kota data regarding housing conditions such as size of the house, number of rooms, built of the house, nature of the house, ventilation facilities, open space in front of the house, etc. have been collected. After tabulation, three levels of QOL, i.e. High, Medium and Low have been identified and the spatial pattern has been depicted in Figure 6.1.

High Level: In Kota 17 wards have been included in high level of QOL based on housing conditions. These wards are 15, 20, 21, 23, 34, 35, 39, 40, 41, 49, 50, 51, 54, 56, 57, 59 and 60. Important colonies included in this category are Civil Lines, Military Area, Vallabhbari, Gumanpura, Talwandi, Jawahar Nagar, Dadabari, Indra Vihar, R.K. Puram, Talwandi, Vigyan Nagar, Srinathpuram, etc. There are certain small colonies in between middle level which are having high QOL such as in North Kota, M.B.S. Nagar, Jai Hind Nagar, Gopal Vihar, Kehsav Nagar, etc. In these areas size of house varies from 2,000 sq. ft to 4,000 sq. ft. Houses are well built and in good design. Most of the houses are having three rooms with lobby, kitchen, more than two bathrooms and well ventilated. Every house has waste water

Figure 6.1 *Kota: Quality of Life Based on Housing Conditions*

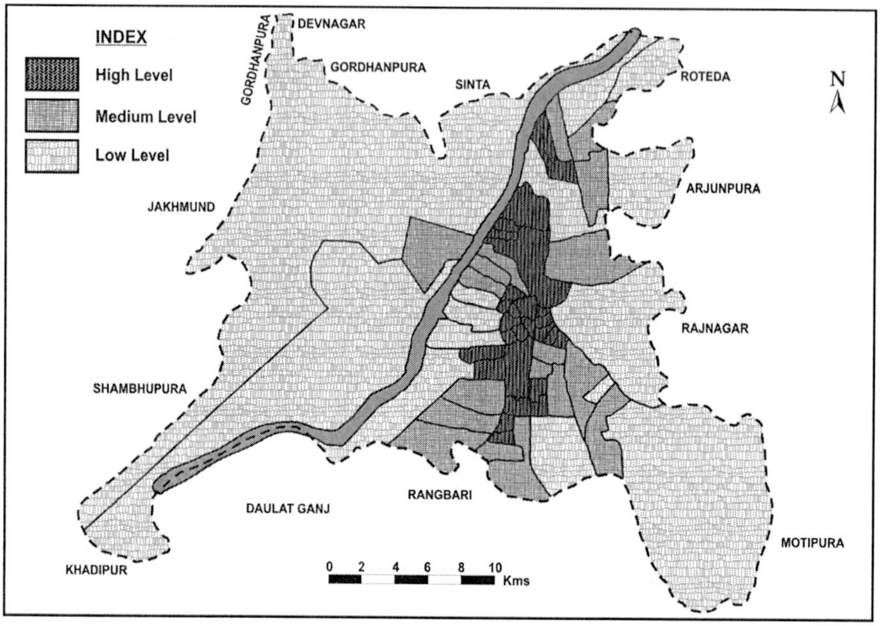

drainage system and most of the houses are having open space in front of them. Flat culture is also now developing in Kota. In Bhimganj Mandi area and in Gumanpura area several flats have been constructed. These flats are also having good housing conditions. Several flats are now developing at Baran Road, Jhalawar Road, Bundi Road and also at CAD Circle, Rawatbhata Road and are having good housing facilities.

Medium Level: All 22 wards of the Kota city are having medium level of QOL based on housing conditions. These wards are 4, 5, 6, 12, 14, 16, 19, 24, 25, 26, 27, 28, 29, 30, 36, 37, 38, 42, 43, 55 and 58. Mostly these wards are located in central region of the city. In these areas the size of the house varies from 400 sq. ft. to 1,500 sq. ft. The built of the house in these areas is of mixed type. Nearly 80 per cent houses are *Packka* with more than three rooms in the house. Although some pockets of slums have also developed in Anantpura, Keshavpura, Udhyog Nagar and Chattrapura. Most of the houses are having double storey and having minimum to good facilities. Most of the old Kota, Talwandi, Mahaveer Nagar, Khelri Phatak, Baran Road area are included in this category.

Low Level: In Kota areas of low QOL in terms of housing conditions are widespread throughout the city. It is interesting to note that such areas are located around the city, especially in western part of the city on western side of river Chambal and in whole of eastern part of the city. Such areas are prominent in ward no. 1, 2, 3, 7, 8, 9, 11, 13, 17, 18, 22, 31, 32 and 33. In ward no. 10 (Kansuwa) all the localities are *Kachhi Basti* (slums). In these areas nearly 50 per cent houses are *Kachha* and their area varies from 50 sq. ft to 50 sq. ft. Most of them are unauthorised. In Kota some of the specific localities included in this category are Chandraghata, Tipta, Ram Chandrapura, Kotadi, Develi Arab, Raipura, Borekhera, Bajrang Nagar and areas along river Chambal. The houses are very congested in all these areas and not having proper ventilation facilities. The waste water disposal facilities do not exist in these areas. Most of these areas are disease prone areas of the city.

QOL Based on Waste Disposal Facility

In cities like Kota, waste collection and disposal facility have had a direct impact on spatial pattern of QOL. The quantity of waste created by the people and industries as well as by other commercial establishment is increasing day by day and now has become a main cause of the degradation of environment and have a direct impact on QOL. During fieldwork primary data have been collected regarding facilities of waste collection and its disposal. Whether such facility is available, if yes, what is its periodicity. Three categories of QOL based on waste collection and disposal have been depicted in Figure 6.2.

High Level means those areas, where regular system of waste collection is in occurrence. Because of daily waste collection both domestic and municipal, these areas are better from hygienic point of view and have lesser

Figure 6.2 *Kota: Quality of Life Based on Waste Disposal Facility*

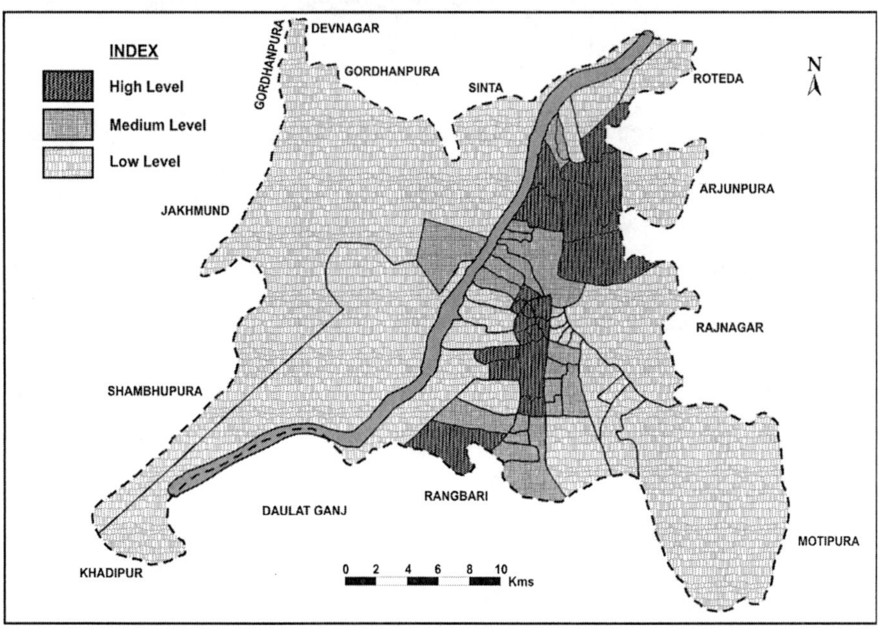

environmental degradation. On the basis of fieldwork, wards included in this category are 4, 5, 12, 13, 14, 15, 21, 22, 23, 34, 35, 40, 49, 50, 51, 54 and 56. The main colonies having high QOL in terms of waste disposal facility are Dadabari, Talwandi, Basant Vihar, Jawahar Nagar, R.K. Puram, Srinath Puram, Civil Lines, M.B.S. Nagar, Army Area, Vallabh Nagar, etc.

Medium Level of QOL in terms of waste disposal facility are in 19 wards. In these areas waste is collected twice in a week, although on paper daily collection of waste has been recorded. The areas included in this category are Ganeshpura, Nayapura, Nehru Nagar, Ladpura, Rampura, Bajaj Khana, Khai Road, Karbala, Vigyan Nagar, Mahaveer Nagar, Keshavpura, Patan Pole, Sripura, Suraj Pole, Kethunipole, Housing Board of Kunhari, etc. Because of irregular waste collection, at several places one can see heaps of garbage, which creates environmental problems.

Low Level of QOL areas in terms of waste collection and disposal are those in which only occureal clearance of waste is done, sometimes in 15 days or in a month. These areas are widespread in whole western and eastern region of the city, as well as in north-eastern and on southern bank part of eastern bank of river Chambal. The main areas in low level QOL are Kala Talab, Purohitji ki Tapari, Areas near Rangpur Bridge, Gavadi, Dostpura, Harijan Basti, D.C.M. Nagar, Prem Nagar, Indraprastha Industrial Nagar, Kunhari, Tipta, Ramchandrapura, Anantpura, Kotari, etc. In several other pockets small areas of this category can be seen.

QOL Based on Internal Drainage Facility

Internal drainage system or sewerage system is an important aspect of environmental quality and indirectly of QOL. The waste water both domestic and industrial should be drained properly in order to avoid its adverse effect. In Kota city some areas are having good internal drainage system and other areas are not having proper system, accordingly on the basis of field survey three levels of high, medium and low have been identified as shown in Figure 6.3.

High Level internal drainage facilities exist in ward no. 5, 12, 13, 16, 23, 34, 35, 38, 40, 41, 54, 56, 57, 59 and 60. In these areas sewerage system is proper therefore waste water is drained out to main outlet and there is no stagnation of water in these areas. These areas are same as described above with reference to waste disposal facility.

Medium Level of internal drainage system is prevalent in ward no. 3, 4, 15, 19, 20, 21, 22, 24, 25, 27, 28, 29, 30, 36, 37, 39, 42, 55 and 58. In these wards 40 to 50 per cent area is having proper drainage and remaining is not. In absence of regular clearance drains often remained choked. This is also due to the residents of that area who used to throw all types of waste in the drains.

Figure 6.3 *Kota: Quality of Life Based on Internal Drainage Facility*

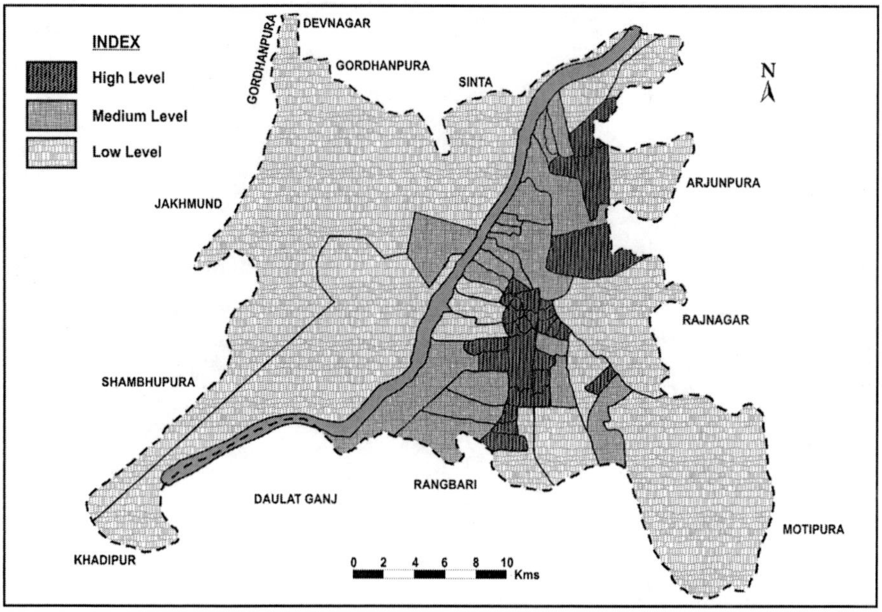

Low Level – A large area of Kota city is not having proper drainage system. These areas are located in whole of the western and eastern part of the city as well as in some southern parts as shown in Figure 6.3. Mostly these areas are either *Kachhi Basti* area or industrial area. The absence of internal drainage is a main cause of low QOL in these areas.

QOL Based on Nature of Roads

Nature of roads means type of streets within the city, i.e. cement concentrate roads, painted roads (B.T. roads), graveled and fair whether roads. The width of the road has also been taken into consideration. Taking into consideration the nature of roads, three levels, i.e. high, medium and low have been determined and depicted in Figure 6.4.

High Level QOL based on roads are found in 26 wards of the city. Most of the new Kota and Military area and Civil Lines are having good road system. The main roads in these areas are having one way system and their width is 60 to 80 feet. The inner roads are also in good condition. Thus, roads in these areas provide not only better transportation facility but also help in reduction of pollution.

Medium Level of roads are in those areas where 60 to 70 per cent roads are in better condition, while others are not properly maintained. Nayapura, Bhimganj Mandi area, Indraprastha Industrial area, Prem Nagar, Rampura, Vigyan Nagar, Kishorpura, Sripura, Chawani area, Dhanmandi, Rangbari area, etc. are in this category and having medium QOL.

Figure 6.4 *Kota: Quality of Life Based on Nature of Roads*

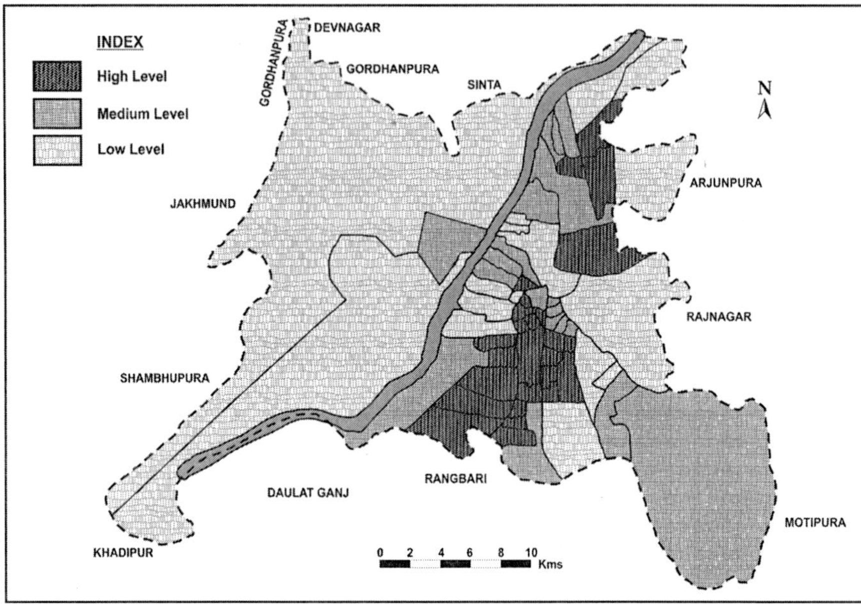

Low Level of roads are the characteristic of slum area (*Kachhi Basti*) and unauthorised colonies, which are widespread as shown in Figure 6.4. Some of the old city areas such as Ladpura, Karbala, Bajaj Khana, Rampura, Rampura (internal side), Sripura, Tipta, Kethunipole, etc. are also not having proper road system. The width of inner roads in many parts of the old city is 4 to 6 feet only. While in *Kachhi Bastis* only graveled roads exist, which becomes drains during rainy season.

QOL Based on Drinking Water Facility

Drinking water is a prime requisite for human beings, therefore, an indicator of QOL, because in absence of water one cannot imagine better QOL. In comparison to other cities of Rajasthan, Kota city is in advantageous position because it has river Chambal as a source of water. Till 1980, Kota was the only city of the State having 24 hours water supply. But with the growth of population and also demand of water in other sectors this practice has not only been stopped but now some parts of the city are facing water crisis (Figure 6.5).

The wards having good supply of drinking water are 5, 6, 12, 14, 16, 23, 34, 35, 38, 40, 41, 42, 49, 50, 51, 53, 55, 56, 57, 58, 59 and 60. These areas are in high level category. Another 17 wards are in medium level category, but supply of water is quite satisfactory in these areas. Other areas are having irregular supply of water. The areas facing problem of drinking water in Kota

Figure 6.5 *Kota: Quality of Life Based on Drinking Water Facility*

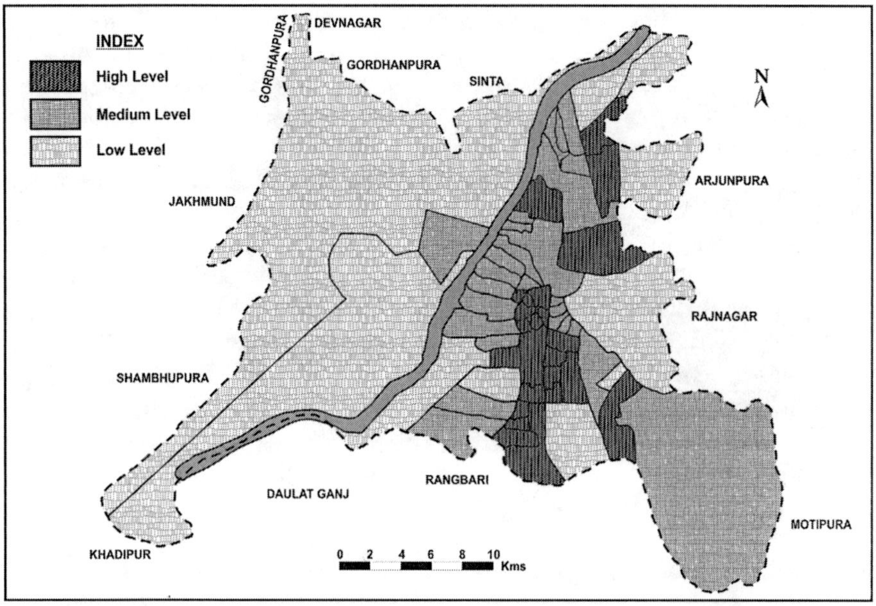

city are Anantpura, Barad Basti, Talav Gavau Chota Barada, Subhash Nagar, Gobaria Bawari, Volbe Yojana, Talab Basti and Deendayal Upadhaya Nagar. The areas located on western side of river Chambal are also having problem of drinking water. The irregular supply of water often creates problems for the residents of that area, in other words it directly affects the QOL, i.e. better supply of drinking water, better the QOL and vice versa.

Spatial Pattern of QOL in Bikaner

Bikaner is a historical city of Rajasthan and located in the arid region of the State. The city was founded in 1488 AD by Rao Bikaji. The city was built in 1700 AD. In Gazetteer of Bikaner, Captain P.W. Powlett described the conditions of Bikaner city in 1874 AD. He narrates that Bikaner looked like a great and magnificent city having a fine wall surrounded by many towers and crowned with the usual Indian battlements. Some high houses and temples rising above the rampart and the striking outline of the lofty past add to the impressive appearance of place. It was only after 1900 AD the city has started its growth outside the city wall and after 1950 several colonies have developed outside the city wall. Now Bikaner city like many other cities of Rajasthan is having old city within city wall and a new city outside the walled city. Bikaner ranks fourth among cities of Rajasthan and according to 2001 census its total population was 5,29,007.

QOL Based on Housing Conditions

Like health care, housing is fundamental for most people. Information on housing is critically important in the locational decision making process. The quality of house available not only affects comfort and convenience, but has a direct bearing on physical health. There are places and public buildings as well as poor huts. The main concern is with the common man's house which portrays the local natural conditions and socio-economic characteristics of the urban society.

In Bikaner city, the layout of houses varies with the economic or social class of the household. According to the facilities available in the houses, the city has been divided into three areas as depicted in Figure 6.6.

(i) *Lower Class Area (Lower Class Houses)*: In Bikaner city, the following areas have predominantly lower class houses. Amar Singhpura, Bangla Nagar, Bhutton ka Bas, Gopeshwar Basti, Harijan Basti, Chhoti Guwad, Indira Colony, Nathusar Harijan Basti, Pratap Basti, Punjab Giran, Railway Tank ke Pass (Sarvodaya) Sangalpura, Subhashpura, Survodayan Basti, (Narsingh Sagar Talab), etc.

Figure 6.6 *Bikaner: Quality of Life Based on Housing Conditions*

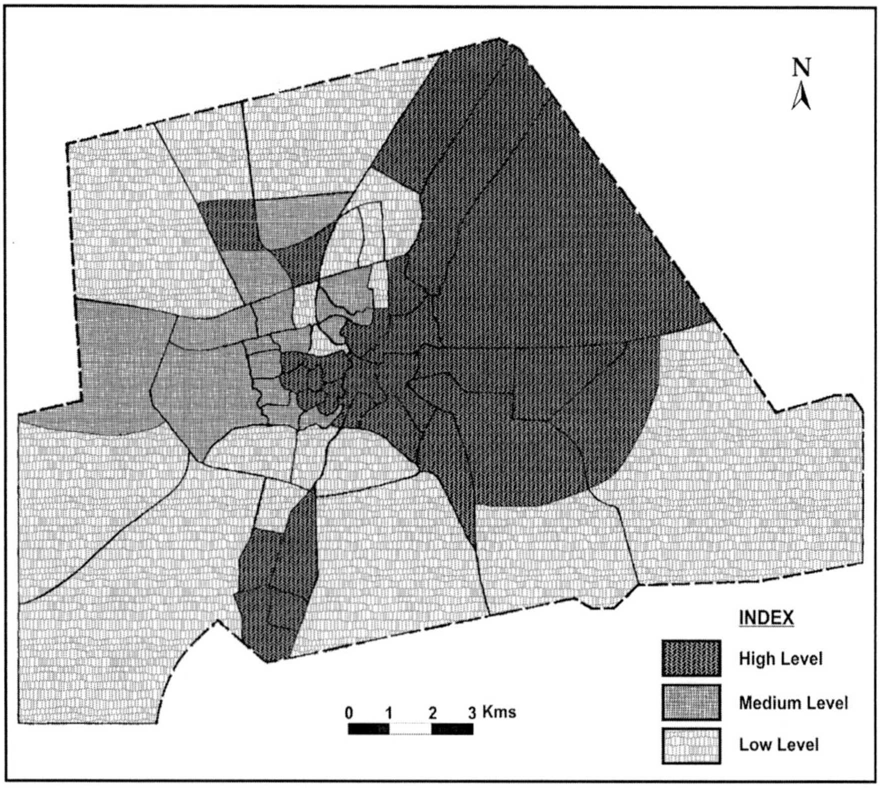

The characteristics of the houses in these areas are (i) The houses are generally having single or in some cases double rooms. (ii) The walls are not plastered. (iii) There is a platform (*Chabutri*) between the street and the house. (iv) Some houses are having three rooms of small size with separate doors opening into the street.

(ii) *Middle Class Area (Middle Class Houses)*: The middle class houses are dominated in the following areas of Bikaner city. Outside Nathusar Gate, Railway Old Colony, Inside Idgah Bari, Inside Nathusar Gate, Bhatron ka Chowk, Uston ki Bari, Inside Goga Gate, Power House, Pareek Chowk, Acharya Chowk, Benisar Well, Mohata Chowk, Bara Bazar, etc.

In old city middle class houses are often called '*havelis*' (mansions). These are ornamented houses built by red stone. This town has an abundance of architecturally rich houses. These are mostly having two to three storeys in which many separate families are living. Most of the rooms in these houses are dark and not having ventilation facility. Inner courtyards are always dingy and dirty very often the toilet opens into the street.

(iii) *High Class Area (Upper Class Houses)*: In Bikaner city upper class houses are located in areas of Jainarayan Vyas Colony, R.S.E.B. Quarters, Military area, Narendra, Bhawan, U.I.T. Colony, R.C.P. Colony, Railway New Colony, Sadul Ganj, Civil Lines, Naganeehiji, Pawan Puri Colony, Post and Telegraph Colony, etc.

One category of such houses has been built during medieval period by royal chiefs and their associates. These are known as '*mahals*' or '*havelis*' but their modern counterparts are known as *kothis* or bungalows. Modern houses have been built from health point of view having proper space for lawns, open space and ventilation facilities, parking space, etc. These houses have separate rooms for kitchen, dining, stores and living. The inner courtyard is well cemented or covered with stone slabs. It may have a front garden as well as kitchen garden, the former usually used for sitting during evening. There are also some modern flats for upper class families. They are not as spacious as bungalows but they are well furnished. Most of them have four to five rooms, well ventilated with high walls, with separate kitchen, bath and toilet.

Besides these houses, in Bikaner city new colonies of modern houses for the working class have been developed by railway, regional offices and R.C.P., Railway, P&T and R.C.P.

QOL Based on Street Conditions

Bikaner city is normal in terms of road transportation facilities which constitute an important part of the environment. But these facilities vary in

Figure 6.7 *Bikaner: Quality of Life Based on Street Conditions*

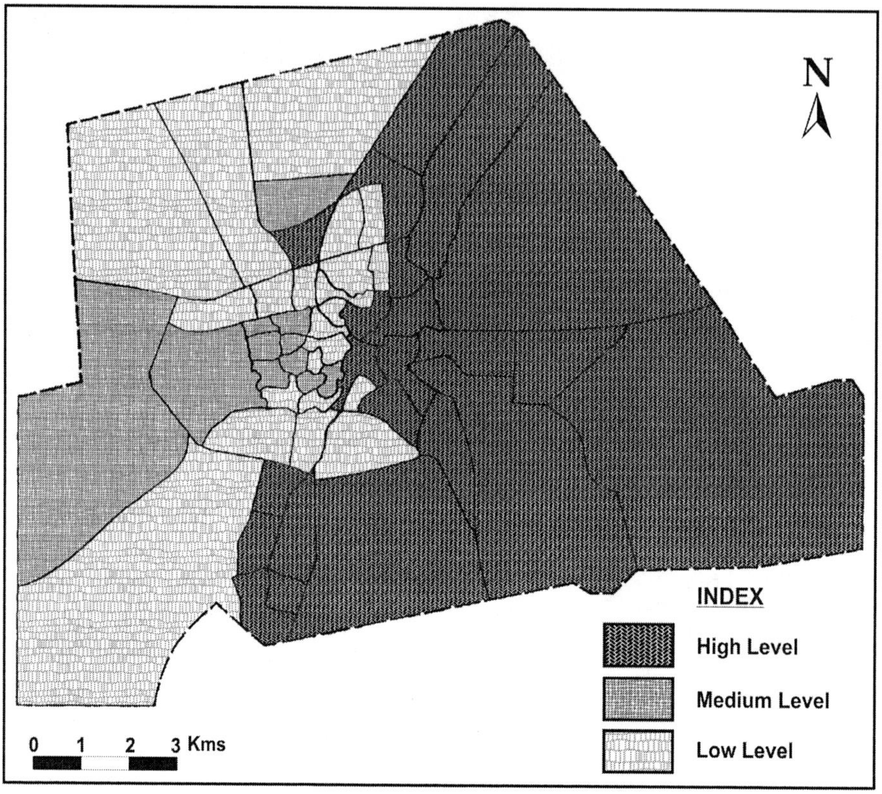

various parts of the city. The total road length within the city is 208 km. Out of this about six per cent are kachha roads (unmetalled). QOL in terms of street conditions is identified as high, middle and lower and has been depicted in Figure 6.7.

(i) *Lower Class Area*: In Bikaner city the worst road conditions are prevalent in ward no. 1, 13, 21, 22, 26, 44 and 52. In these areas roads are unmetalled and kachha. Most of the roads are in the form of narrow lanes. In the rainy season these localities become very dirty due to filthy water. This is more because of unplanned growth of the localities and most of them are unauthorised.

(ii) *Middle Class Area*: These areas include ward no. 2, 28, 32, 33, 48 and 49. The roads of these areas are metallic but narrow. The situation is more critical in the inner area of walled city. In this area due to congestion and building extension even improvement of road system is impossible and the problem of parking is serious. Traffic jams in the city centre particularly in the market area of Bara Bazar is a common experience.

(iii) *High Class Area*: The areas having good conditions of roads in Bikaner city include ward no. 7, 8, 9, 10, 12, 23, 24 and 29. The roads of these areas are metallic, straight and wide. In these areas, the problems of parking and traffic congestion are not serious in comparison to the other areas of the city. Therefore, these areas depict high standard of QOL. These areas include J.N. Vyas Colony, Sadulganj, U.I.T. Colony, Railway Colony, etc.

Quality of Life Based on Drinking Water Facilities

Water is an essential and the most important fundamental physiological need of man. It also meets the requirement of industrial sector. Before the construction of overhead tanks, wells were the only means of water supply. Despite the best efforts of the Public Health Engineering Department in Bikaner many areas are still not receiving proper drinking water. The water supply system of Bikaner city is not well developed. It has no uniformity in all parts of the city. The shortage of water in walled city areas is mainly due to dense population. The supply of water in these areas has become difficult and expensive because of their higher elevation. The spatial pattern of QOL in terms of drinking water supply has been depicted in Figure 6.8.

Figure 6.8 *Bikaner: Quality of Life Based on Drinking Water Facilities*

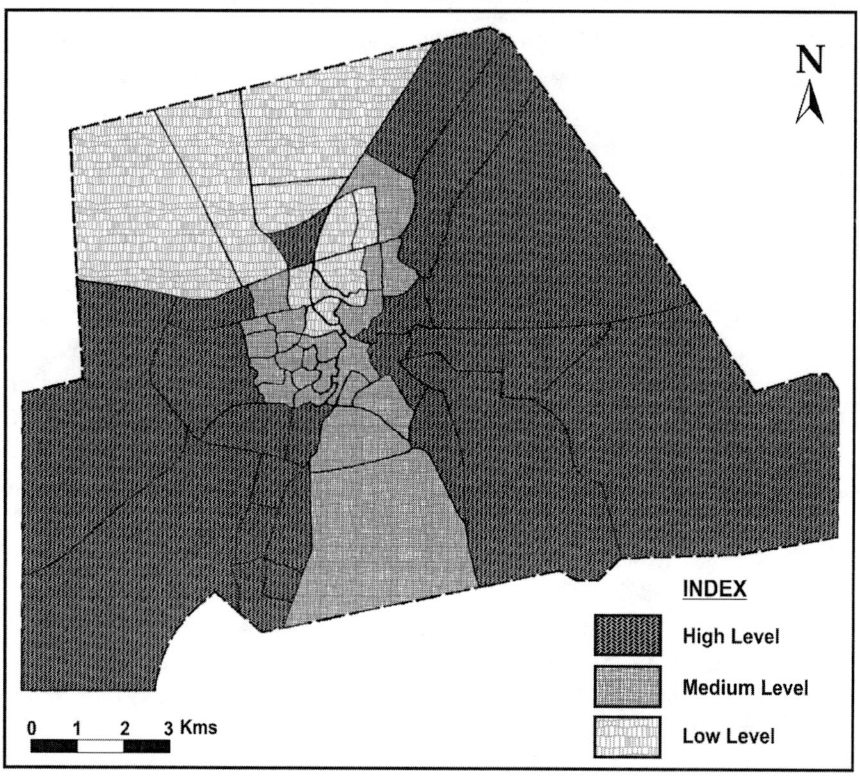

(i) *Lower Class Area*: In Bikaner city Banglanagar, Amarsinghpura, Subhashpura, Rampura Harijan Basti, Bhuton ka Bas, Pratap Basti, etc. are the areas which do not get water even for domestic purpose. The water supply is not satisfactory in these areas since some of these areas lie far away from the source of water supply. People of these areas do not have domestic water taps and they depend on public supply and thus have inadequate water supply. These areas are having lower QOL.

(ii) *Middle Class Area*: The areas inside the city wall do not get adequate water supply mainly due to dense population and elevated location. In many higher areas of the city like Teliwara and Gangashahar road supply of water becomes difficult as they require higher water pressure. Some areas of the city do not get adequate water supply as they are far away from the main distribution line. These areas are classified as areas of middle QOL.

(iii) *High Class Area*: In these areas water supply is regular and adequate. These areas are classified as areas of high QOL in terms of drinking water supply. In these areas density of population is lesser. The main areas included in this category are J.N. Vyas Colony, Sadulganj, U.I.T. Colony, Railway Colony, etc.

Quality of Life Based on Internal Drainage Facilities

In Bikaner city, there is a network of internal drainage for draining the domestic dirty water. These drains are linked with Sursagar from Sursagar to Ballabh Garden. Some drains of the city are linked with Sujandesar via Sri Ramsar Gangashahar and Bhinasar. The three categories, i.e. lower, middle and high in terms of city drainage facilities have been depicted in Figure 6.9.

(i) *Lower Class Area*: The areas not having proper drainage system are Banglangar, Meghwal and Harijan Mohalla, Gangashahar, Bhinasar, Sangalpura, Rampura, Harijan Basti, etc. All these localities also have rural impact. The dirty water used to flow in the mohallas. Thus in these localities epidemics and diseases are common. Some areas like old Ginani, near Station road and Collectorate are low lying areas, these areas have drainage system. But during rainy season dirty water overflows in the same areas instead of the drains. Sometimes water enters into the houses. These areas are classified as low class areas of the city.

(ii) *Middle Class Area*: Although, the areas inside the city wall have drainage system but the residents of all mohallas dump their garbage, polythene bags, etc. in open drains which chokes the flow. Some parts of the city have defective drainage system due to lack of gradient and shallowness. Thus their dirty water overflows in the streets instead of

Figure 6.9 *Bikaner: Quality of Life Based on Internal Drainage Facilities*

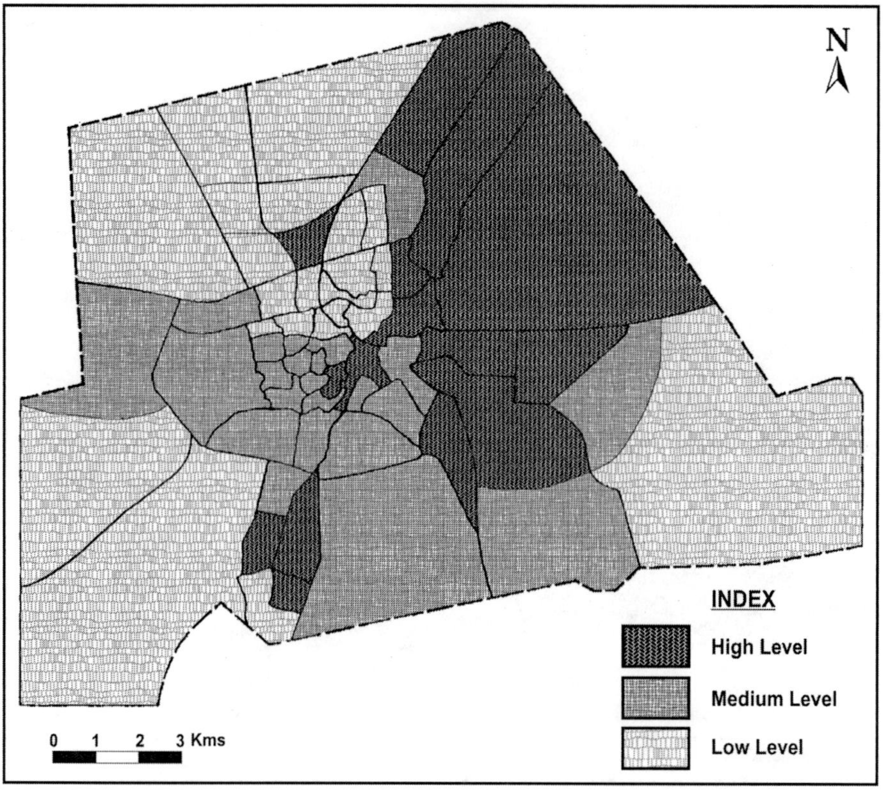

the drains. This overflow of dirty water spreads dirty material on the roads which is unhygienic. It also damages the roads. These areas are classified as middle class areas of the city.

(iii) *High Class Area*: J.N. Vyas Colony, Sadulganj, U.I.T. Colony and Railway Colony, etc. are the high class areas of the city because these areas have a proper drainage system. The drains are clean and dirty water flows in the drains. These areas are the examples of good urban planning. These areas represent high QOL.

Quality of Life Based on Waste Disposal Facilities

The waste contains 'solid wastes', i.e. garbage (food wastes), rubbish (paper, plastic, wood, metal, throw-away containers, glass), demolition products (bricks, masonry, pipes), sewage treatment residue (sludge and solids from the coarse screening of domestic sewage), etc. In Bikaner city disposal of waste also presents a problem. Heaps of garbage in residential localities is a familiar sight. Human excreta and piles of various waste at

Figure 6.10 *Bikaner: Quality of Life Based on Waste Disposal Facilities*

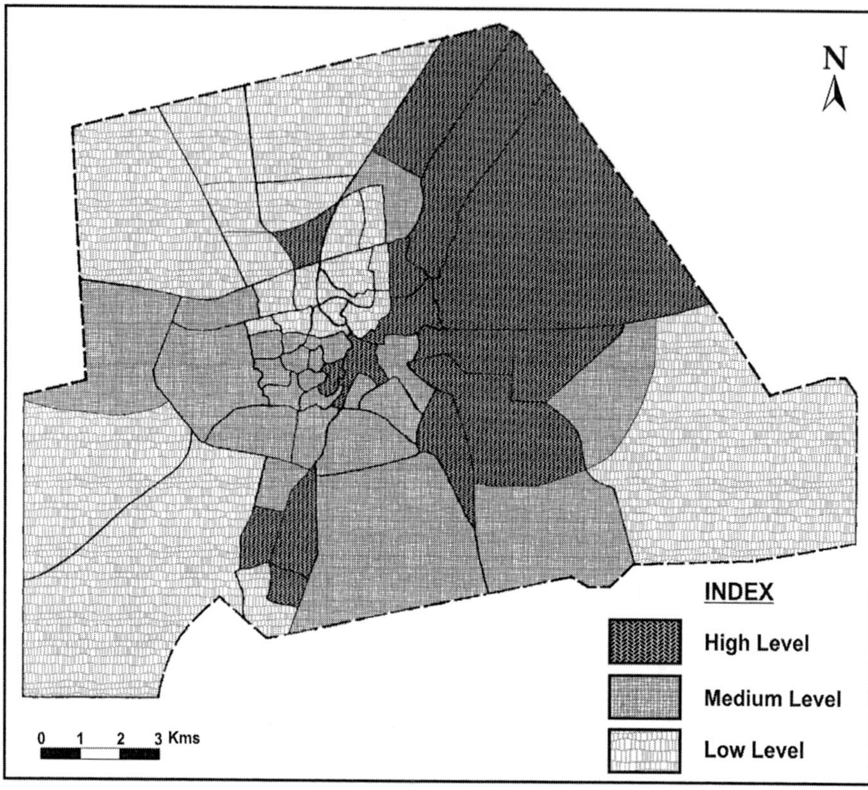

roadside is a common sight. The spatial distribution analysis of the QOL on the basis of waste disposal facilities has been depicted in Figure 6.10.

(i) *Low Class Area*: In Bikaner city the following areas are not having proper waste disposal facilities. Bangla Nagar, Rampura Harijan Basti, Harijan Basti, Button ka Bas, Chooti Guawad, Bari Guawad, outside Guga Gate, etc.

In these areas sewerage and sanitation facilities are unsatisfactory. Stagnant pools of water and flooding of low lying residential areas by rain water are a common sight. Drainage system is non-existent in the slum areas as well as in old part of the city. Heaps of garbage in residential areas is a familiar sight. Septic tanks are rare.

(ii) *Middle Class Area*: The middle class areas in terms of waste disposal are Acharya Ke Chowk, Benisar Bari, Teliwara Murli Dhar Vyas Colony, Bhatron ka Chowk, Joshiwara, etc. In these areas although drainage system is developed but the old 'nallahs' have either been blocked or out of use. So the slushy water spilling out of every household stagnates on the roads and lanes and has become the breeding sites of mosquitoes.

Disposal of household refuse and garbage is not regular. Heaps of garbage in residential areas remain for several days. The situation of these areas is also unhygienic to the residents.

(iii) *High Class Area*: The high class areas in terms of waste disposal are J.N. Vyas Colony, Sadul Ganj, Military area, Narendra Bhawan, R.C.P. Colony, U.I.T. Colony, Naganeechiji, Railway New Colony, etc. These areas have better waste disposal facilities in Bikaner. The sewerage and sanitation facilities are good in these areas, therefore, these areas have been included in high QOL areas of the Bikaner city.

Spatial Pattern of QOL in Sriganganagar

Sriganganagar (29°56'N, 73°52'E) is an important city of northern Rajasthan. It is one of the important strategic town having a *'tri junction'* location at the border of Rajasthan and Punjab States and close to the international border of Pakistan. It is situated midst the agricultural region irrigated mostly by Gang canal. Sriganganagar is a newly developed town of Rajasthan. Prior to 1927 a small hamlet known as 'Dhani' of 'Ramu Jat' was there, which thereafter called Ramnagar. The development of present town was started with the advent of Gang Canal in 1927. After inauguration of Gang Canal in 1927, the Maharaja Ganga Singh named it as Sriganganagar. The foundation of the new planned town was laid down by the Maharaja in 1931. It was a small urban centre in 1931 with a population of 9,943. But with the rapid growth of population, the town has also developed at a fast rate. After independence the growth of population and town was rapid because a large number of refugees were settled in this town. Now Sriganganagar ranks ninth in the cities of Rajasthan with a population of 2,22,833 in 2001. It is interesting to study the QOL in Sriganganagar because it represents a newly developed town of the State and a planned town. The study of the QOL in Sriganganagar is based on intensive fieldwork. The spatial pattern of the QOL based on selected variables is as follows:

QOL Based on Housing Conditions

The housing conditions, i.e. size and built of the house, number of rooms, open space in front of the house, ventilation facilities and number of family members living in a house are the best indicators of the QOL. On the basis of fieldwork following three categories, i.e. high, medium and low have been identified and their spatial pattern has been depicted in Figure 6.11.

High QOL: In Sriganganagar high QOL areas are Jawahar Nagar, Sukhadia Nagar, Agarsen Nagar and Civil Lines. In these areas all the houses are well built and having more than three living rooms with good ventilation facilities. Nearly 70 per cent houses are also having small open space.

Medium QOL: A large area of Sriganganagar is under the medium level of QOL in terms of housing conditions. The central part of the city known

Figure 6.11 *Sriganganagar: Quality of Life Based on Housing Conditions*

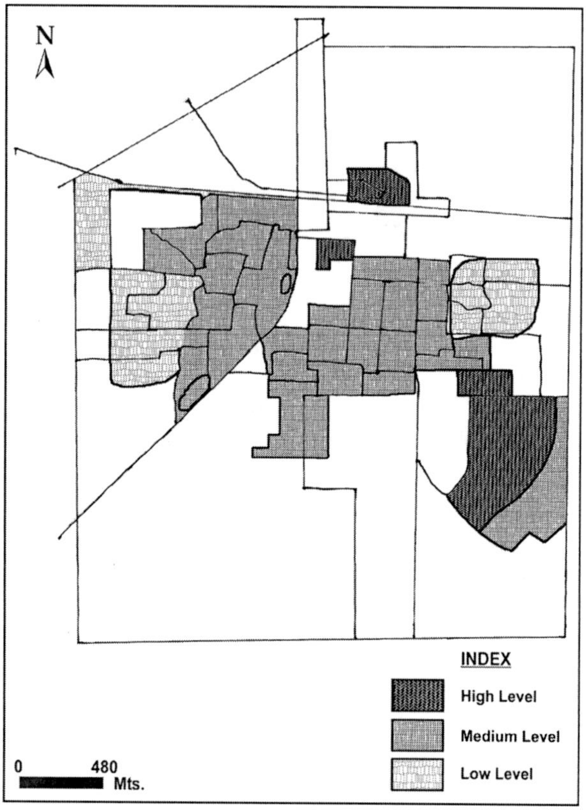

as Block Area (H, P, N, F, B Blocks), Vinobha Basti, ward no. 12, 13, 14, 17 of Purani Abadi, area near Tarachand Vatika, Khalsa Nagar, Sadbhawna Nagar, Colonies developed along NH–15 are having medium QOL in terms of housing conditions. These areas are equally good, but congested and not having open space in front of houses. The built of the house in these areas is good, but size of the rooms are lesser than high quality houses. In these areas 65 per cent houses are having good ventilation facilities, 20 per cent limited and 15 per cent are not having proper facilities.

Lower QOL: The areas having low QOL in Sriganganagar are scattered around the outer side of the city. These areas include Sugar Mill Colony, Guru Nanak Basti, Areas along Karanpur road, Dev Nagar, Arjun Nagar, JCT Labour Colony, Western part of Purani Abadi, Kedar Nagar, Subhash Nagar, Muslim Mohalla, Shyam Nagar, Khatik Mohalla, Sampat Basti, etc. In these areas the size of the house varies from 50 sq. ft to 150 sq. ft. About 42 per cent houses are having single room without any ventilation facility. About 30 per cent houses are Kachha in nature and not having water connection. Most of the people living in these areas are of BPL category.

QOL Based on Internal Drainage Facilities

The internal drainage facility is meant for the disposal of waste water both domestic and municipal. In Sriganganagar internal drainage is a great problem in many areas of the city. This has become more because the land is level and not having slope.

The only area having good internal drainage facility is the Civil Line area, which can be considered as a high class area. About 60 per cent area of Sriganganagar has been included in middle class area in terms of internal drainage facility. In these areas following residential areas are included Whole block area, Sukhadia Nagar, Jawahar Nagar, Housing Board colony, Goshala road colony, Vinoba Basti, Satia Colony, Khalsa Nagar, Kunj Vihar, Eastern part of Purani Abadi, etc. In these areas drainage is there but drains are shallow and their clearance is also not regular.

On the other hand several colonies are either not having drainage facility or having very limited facility. In several areas waste water spreads

Figure 6.12 *Sriganganagar: Quality of Life Based on Internal Drainage Facilities*

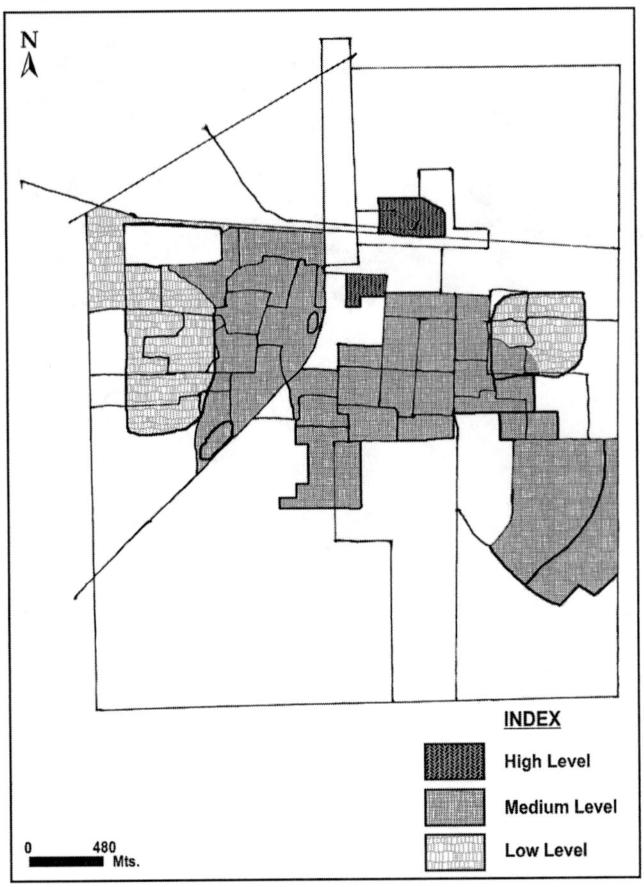

either on streets or remains stagnant in some lower area. A large area near sugar mill has become a collection centre of waste water, thus, a main cause of environmental degradation in this area. Similarly waste water of several colonies is drained near Sadbhawana Nagar and this water now has become a great problem of environmental degradation for the whole city because there is no facility for its treatment. The other areas included in lower QOL in terms of internal drainage facility are JCT Mill Colony, Karanpur Basti, Western Part of Purani Abadi, Bharat Nagar, Dev Nagar Labour Colony, Shakti Nagar, Shyam Nagar, Kedar Nagar, Subhash Nagar, Guru Nanak Basti, Bapu Nagar, Ravidas Nagar, Ashok Nagar, etc. (Figure 6.12).

QOL Based on Drinking Water Facilities

In Sriganganagar conditions of drinking water supply is better in comparison to many other cities of Rajasthan. But within the city the condition of availability of drinking water varies from area to area, which affects the city QOL.

Figure 6.13 *Sriganganagar: Quality of Life Based on Drinking Water Facilities*

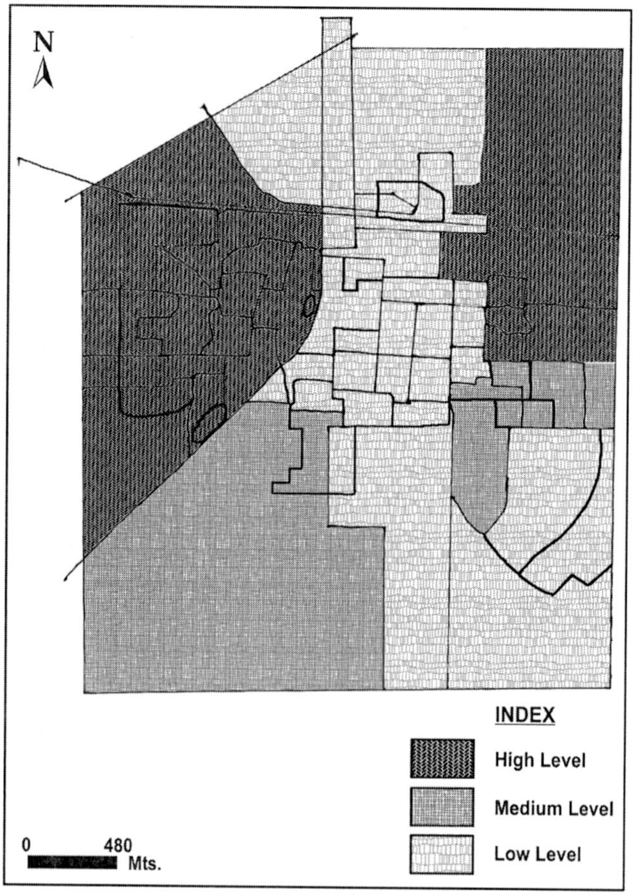

In Sriganganagar *high class areas* in terms of drinking water supply are Civil Lines, Sukhadia Nagar, Jawahar Nagar and Housing Board Colony. These areas are having regular water supply twice a day and all the houses located in these areas are having direct water connection from government water supply system (Figure 6.13).

The *medium class areas* in terms of drinking water supply are quite widespread in the city and covers two-third part of the city. The important areas included in this category are all the Block Areas, Goshala Area, Vinoba Basti, Satia Colony, Eastern half of Purani Abadi, Khalsa Nagar, Colonies along Padampur and Hanumangarh Road, etc. In these areas also water supply is regular but time is less than high class area.

The *Low QOL areas* in terms of drinking water are Western Part of Purani Abadi, Dev Nagar, Arjun Nagar, Kedar Nagar, Subhas Nagar, Shyam Nagar, Labour Colony, Basant Vihar, Gurunanak Basti, Indra Colony, Bapu Nagar, Ravidas Nagar, etc. These areas are having irregular supply of water. Nearly 68 per cent houses in these colonies are not having direct water connection and depend upon public water tap or on hand pumps for their drinking water.

QOL Based on Waste Disposal Facility and Street Conditions

In Sriganganagar areas of waste disposal facility and street conditions are similar. Generally areas having good waste collection and disposal facilities are also having better roads.

The *high class areas* in terms of waste collection and better streets are Block area, Jawahar Nagar, Sukhadia Nagar, Civil Lines, Vinoba Basti, etc. Sriganganagar is a planned town, therefore, having better roads. In these areas domestic and municipal waste is collected regularly. The roads are metalled and clean. While a large area of the city is having *medium class* waste disposal facility. Area along Goshala road, Birbal Chowk, Padampura Road, Bapu Nagar as well as newly developed colonies along Hanumangarh road, Khalsa Nagar, Ashok Nagar, Kunj Vihar, etc. have been included. The waste collection is done once or twice in a week. Although street conditions are satisfactory.

The *lower class areas* of Sriganganagar are widespread around the city. In this category whole of Purani Abadi has been included. Purani Abadi is an unplanned area of the town and street conditions are also not satisfactory and waste disposal facilities are also poor. Other areas having lower class are Sugar Mill Basti, Guru Nanak Nagar, Ashok Nagar, Ravidas Nagar, Karanpur Basti, Dev Nagar, Arjun Nagar, Bharat Nagar, Shakti Nagar, Labour Colony, Kedar Nagar, Subhash Nagar, Shyam Nagar, etc. In these areas waste collection facility is not satisfactory and heaps of waste can be seen here and there. These areas are the disease prone areas of the Sriganganagar and having low QOL.

Spatial Pattern of QOL in Bharatpur

Bharatpur is an important city of eastern Rajasthan having population of 2,05,235 in 2001. It is a fast growing city and has recorded high decennial growth of population, specially during 1971–81 and 1981–91 where the percentage growth rate during these decades were 50.68 and 49.02 per cent respectively. Although, during 1991–2001 it was reduced to 30.82 per cent. The city has expanded at a fast rate in terms of residential areas as well as a commercial centre. The impact of the growth of infrastructures or in other words city development can be seen in its various aspects including the QOL. In fact, QOL is a reflection of overall development, both positive and negative. The present analysis of the QOL in Bharatpur city is based on detailed fieldwork. The QOL in Bharatpur has been examined on the basis of five selected variables and our analysis reveals the following pattern.

QOL Based on Housing Conditions

On the basis of housing conditions, i.e. nature of house, its size, number of rooms, open space in front of the house, three categories of QOL in Bharatpur has been identified, i.e. areas of lower, medium and high. The same has been depicted in Figure 6.14.

Figure 6.14 *Bharatpur: Quality of Life Based on Housing Conditions*

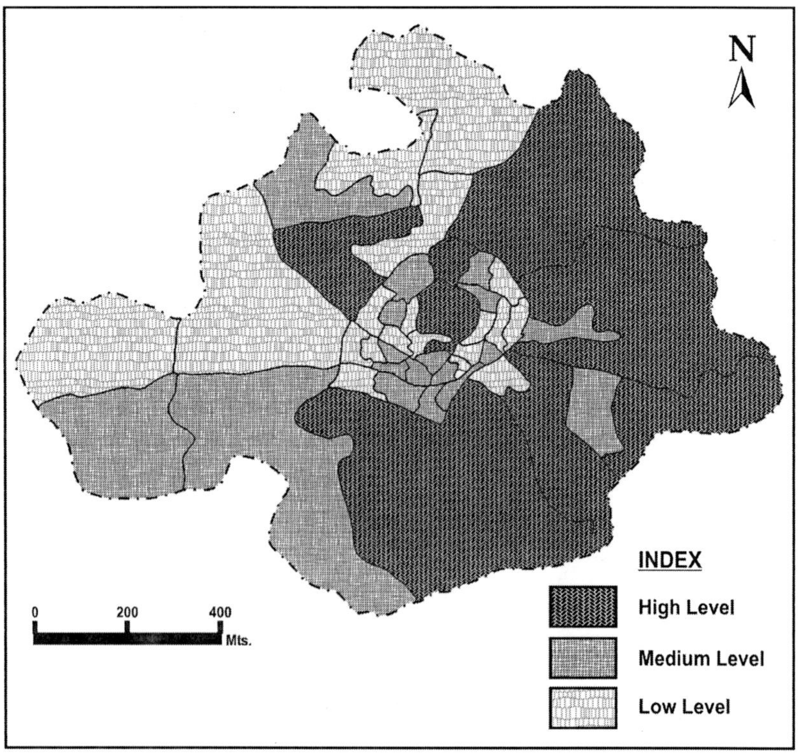

(i) *Lower Order QOL*: In Bharatpur out of total 45 wards, 17 wards are in the category of lower order. These wards are 1, 2, 6, 7, 8, 17, 20, 27, 28, 30, 31, 32, 34, 39, 41, 42 and 43. In these areas more than 50 per cent houses are *Kachha* and their area is less than 200 sq. ft. In ward 7 (*Dhobi Basti*) 90 per cent houses are *Kachha*. In these areas houses are not having minimum ventilation facilities and most of the houses are having only one or at least two small size rooms for the family of 6 to 10 members. In fact, these areas are mostly slum areas.

(ii) *Medium Order QOL*: Twenty wards have been included in this category, these are 3, 4, 9, 10, 11, 12, 13, 14, 15, 16, 18, 19, 22, 25, 26, 33, 35, 37, 38 and 44. In these areas size of the house varies from 200 to 500 sq. ft in 40 to 60 per cent houses, while others are having 500 to 1,000 sq. ft houses. On an average 10 per cent houses are *Kachha* and remaining 90 per cent are *Packka*. Except Sanjay Nagar and Company Bag Colony all other colonies are unplanned. In these wards most of the houses are of two rooms with kitchen, toilet and also have water and light connection. During rainy season many areas and streets are facing problem of waterlogging. Most of the medium income group of family reside in these areas.

(iii) *High Order QOL*: In this category eight wards, i.e. 5, 21, 23, 24, 29, 36 and 45 have been included. Most of the houses in these areas are in size more than 2,000 sq. ft and their built is also good. In these areas 90 per cent houses are having more than three rooms with other facilities. In Jawahar Nagar, New Civil Line, Krishna Nagar and Ranjeet Nagar nearly 50 per cent houses are double storeyed. In the city Jawahar Nagar, New Civil Line, Rajendra Nagar and Krishna Colony are known as V.I.P. colonies. In these areas QOL is good and their standard of living is also high in comparison to other areas of the city.

QOL Based on Waste Disposal Facilities

The waste disposal facilities are of prime concern for the QOL in every city. The waste, both domestic and municipal is growing day by day and if it is not properly collected and disposed its impact is direct on QOL. On the basis of waste disposal facilities in Bharatpur city three types of areas, i.e. low, medium and high have been identified and depicted in Figure 6.15.

(i) *Low order*: In Bharatpur 15 wards are not having proper waste disposal facilities. These areas are located in eastern, northern and central parts of the city in ward no. 1, 2, 3, 7, 9, 13, 20, 27, 28, 29, 34, 36, 39, 42 and 43. Nearly half of the slums of the city are located in these areas, which are not having waste collection facilities. In these areas one can

Figure 6.15 *Bharatpur: Quality of Life Based Waste Disposal Facilities*

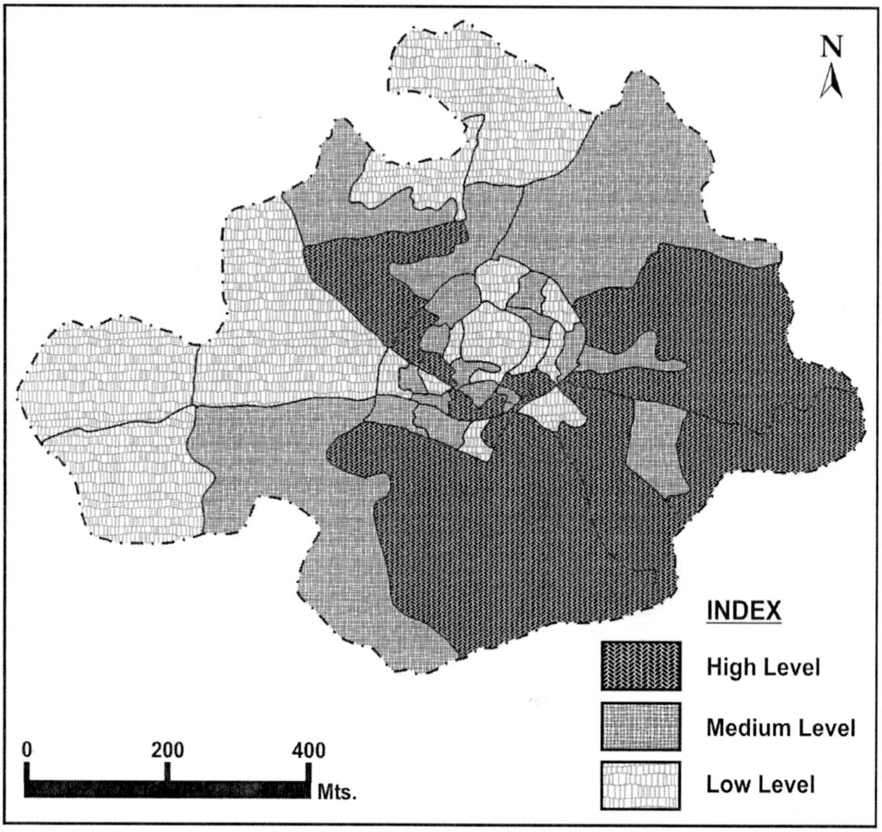

see humps of waste here and there. The industrial areas are also not having proper waste disposal facilities. In many areas the waste is not even collected once in a month. Therefore, these areas are disease prone areas of the city.

(ii) *Middle order*: The middle order areas are located in south-eastern, north-eastern, central and in a pocket of eastern part of the city. In all 18 wards are having limited waste collection facility. In these areas waste is collected mostly once in a week. In some colonies such as Surajmal Nagar, Housing Board the waste is collected twice in a week. In absence of daily collection of waste in several areas waste become a cause of several diseases.

(iii) *High order*: In terms of waste collection and disposal 12 wards are having satisfactory facility. In these areas generally waste is collected daily. The areas located in eastern, southern, south-eastern and some central part of the city have been included in this category.

Figure 6.16 *Bharatpur: Quality of Life Based on Drinking Water Facility*

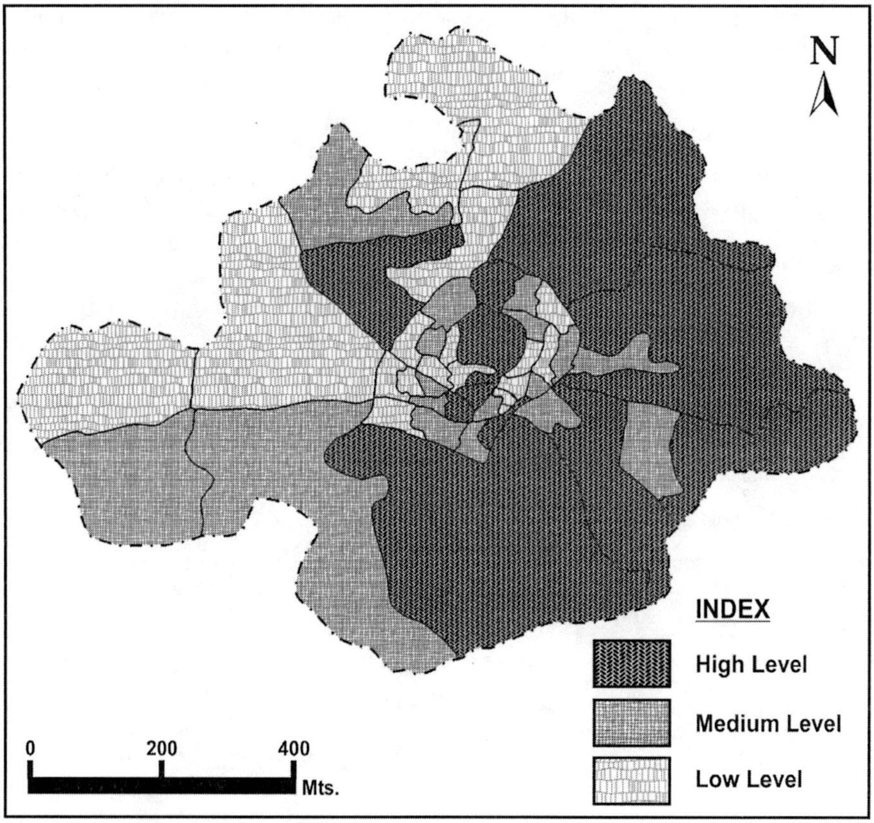

QOL Based on Drinking Water Facility

Drinking water is a minimum requirement for the QOL. The areas having high, medium and low availability of drinking water in Bharatpur has been shown in Figure 6.16. It becomes clear from the figure that 11 wards located in eastern, north-eastern and southern parts of the city are having regular water supply in their homes. These areas are also having high QOL in terms of other variables. On the other hand 16 wards of the city, mostly slum areas are not having water connection in their houses and depend upon public taps or on hand pumps for their drinking water. The regularity of water supply is also not there. The remaining areas of the city are having satisfactory drinking water facilities, thus having medium QOL.

QOL Based on Street Conditions

In Bharatpur street conditions in 9 wards, i.e. 1, 6, 7, 13, 20, 27, 28, and 39 are below the standard. Most of the roads in these areas are still not painted and often waste water flows on the streets. Their condition become worst during rainy season. These areas are having low QOL. While the medium

QOL in terms of street conditions can be seen in 22 wards of the city. In these wards most of the streets are painted but in absence of maintenance their condition is not proper. While in 15 wards of Bharatpur the road conditions are comparatively good.

QOL Based on Internal Drainage Facility

On the basis of fieldwork a map has been prepared (Figure 6.17) showing QOL in terms of drainage facility. In fact, drainage facility for the waste water disposal is a necessary condition for QOL, because in absence of proper drainage polluted water becomes a main cause of several diseases. It becomes clear from the maps that in as many as 15 wards of the city, internal drainage is of low order, while in 16 wards it is medium and in remaining 14 wards it is good. The old city is not having proper drainage, on the other hand the nearly built colonies are having proper drainage facilities. In slum areas there is complete absence of drainage system, thus the QOL in these areas is quite low.

Figure 6.17 *Bharatpur: Quality of Life Based on Internal Drainage Facility*

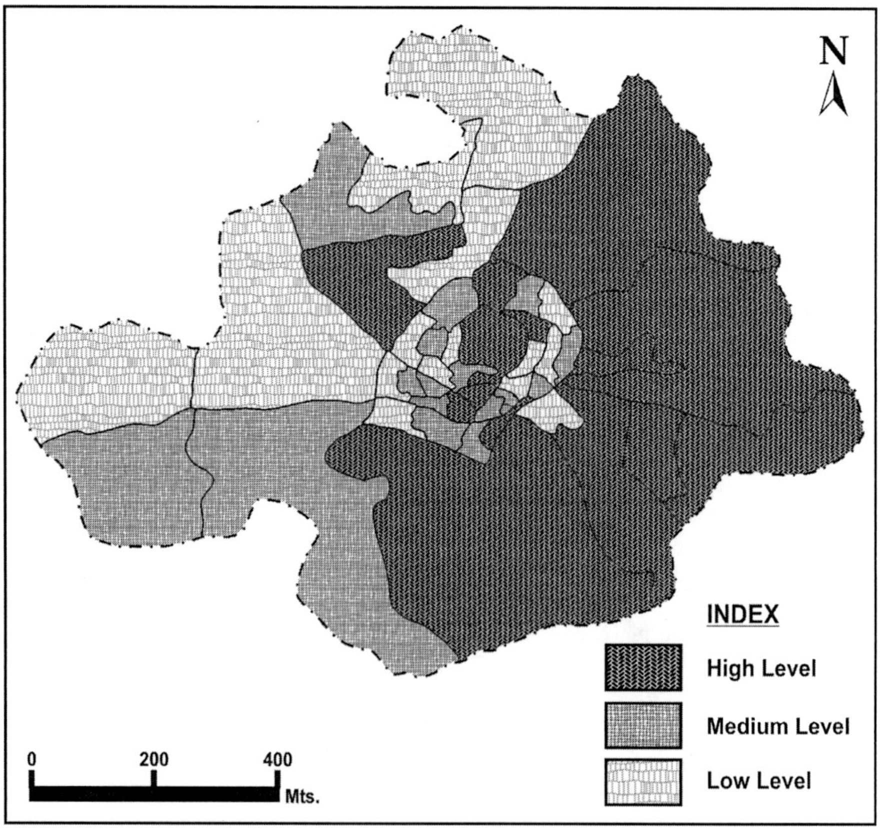

Spatial Pattern of QOL in Tonk

Tonk city is located at the southern bank of river Banas. According to 2001 census, the population of Tonk is 1,35,689. The city is well connected by roads and National Highway No. 12 passes through the city, although Tonk is not having railway facility. Being a district headquarter it is one of the important market centre of the region. Historically also, Tonk is important and during Medieval Period till independence it was ruled by *Nawabs*. In fact, Tonk is an old city, congested and represents a mixed culture, i.e. Hindus and Muslims. Now the city is growing at a fast rate across its old boundaries. With the growth of city several problems have cropped up including degradation of environment, which has already been discussed in the last chapter. The spatial pattern of QOL in Tonk reveals some interesting findings, therefore, Tonk has been selected for sample study. The spatial pattern of QOL in Tonk has been determined on the basis of selected variables, i.e. housing conditions, waste disposal facility, drinking water facility, internal drainage facility and street conditions. The detailed analysis based on fieldwork is as follows:

QOL Based on Housing Conditions

Housing conditions, i.e. built of house, its size, number of rooms and other facilities is a good measure of QOL because it indicates the standard of living and status of the family. Three categories of QOL, i.e. high, medium and low have been determined and depicted in Figure 6.18.

Figure 6.18 *Tonk: Quality of Life Based on Housing Conditions*

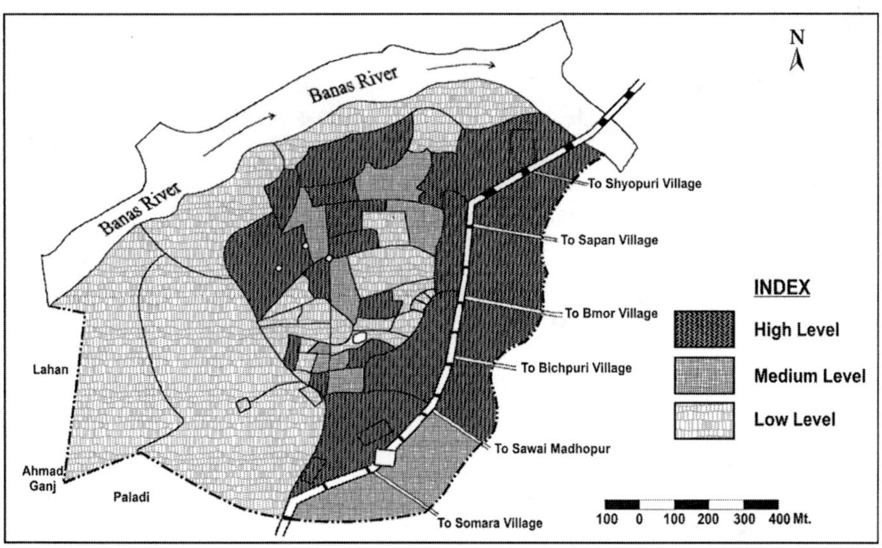

The High Level of QOL in Tonk is limited in only four wards, i.e. ward no. 4, 18, 19 and 21 which includes Civil Line area, Housing Board Colony, Indira Colony and Shastri Nagar. In these areas the size of the house varies from 2,000 to 3,000 sq. ft and all the houses are well built. These houses are having three or more than three rooms as well as kitchen, bathrooms, etc. and about 60 per cent houses also have small open space in front of houses. In Tonk, these areas are having better housing conditions in comparison to other areas, therefore, included in areas of high QOL.

In Medium Level of QOL, 24 wards have been included. In these areas the size of the house varies from 500 to 1,000 sq. ft. In these areas 70 per cent houses are *packka* and 30 per cent *kachha.* Most of these areas are located in Purani Tonk and are mostly unplanned and congested. Although in old houses number of rooms are many, varies from four to ten but their size is small, sometime 10'×8' or even lesser. From health point of view the housing conditions in these areas are not satisfactory.

The Low Level of QOL in Tonk is spread in as many as 18 wards of the city. The whole western, the northern part along Banas river and central part of the city is in low category of QOL. Most of these areas have been declared as Kachhi Basti (Slum Area). In these areas some specific areas are Gole Dungari, Gadda Pahadia, Shargran Pahadia, Qafla Bainwan, Harijan Basti, Ragran Basti, Kassaban, Dhanna Talia, Kumharon ki Chowki, etc. Low roofing, small rooms, no proper ventilation are the characteristics of the houses in this area. The size of the house in these areas varies from 100 sq. ft to 200 sq. ft. Most of the families living in these areas are of lower income group.

QOL Based on Waste Disposal Facility

The waste disposal is a great problem in Tonk city, because the city is unplanned and its Municipal body is not having resources to manage waste collection and its disposal. The situation is worst in 18 wards of the city. These wards are 2, 3, 8, 9, 10, 11, 17, 24, 25, 26, 27, 32, 33, 38, 44 and 45 and are in the low category of QOL based on waste disposal facility. These wards are not having any waste collection facility and only once or twice in a month, the domestic and municipal waste is collected. Medium level of QOL in terms of waste disposal facility is spread in ward no. 2, 4, 5, 6, 9, 14, 18, 20, 24, 28, 30, 32, 38, 39, 40, 41, 42, 43, 44 and 45. In these areas waste is collected once or twice in a week. Most of the people used to throw their domestic waste in front of their house, which creates problems. In Tonk only four wards, i.e. ward no. 4, 18, 19 and 21 are having regular waste collection facility. Such areas are Civil Lines, Housing Board Colony, Shashtri Nagar and Adarsh Nagar Colony and are included in high QOL area of Tonk (Figure 6.19).

Figure 6.19 *Tonk: Quality of Life Based on Waste Disposal Facility*

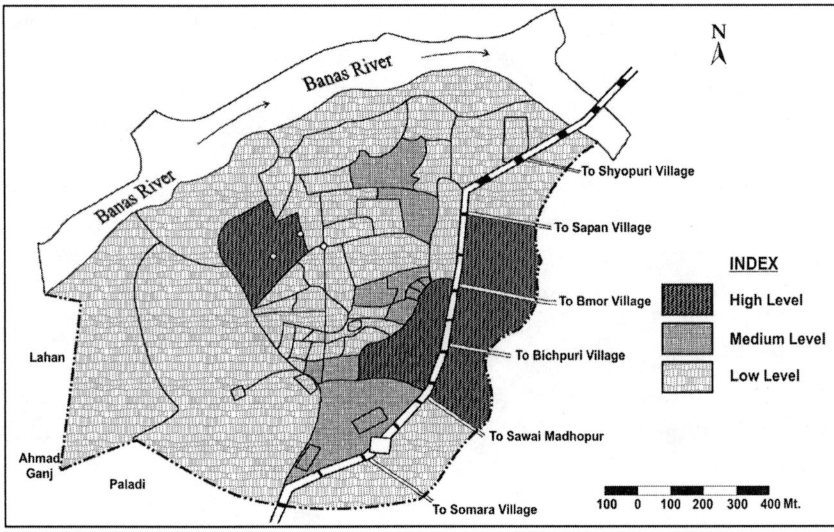

QOL Based on Drinking Water Facility

Although Tonk is located at the bank of river Banas, but nowadays facing drinking water problem. A vast area of Tonk city is under low QOL in terms of drinking water facility (Figure 6.20). It becomes clear from the maps that south-western, southern and northern areas as well as Purana Tonk area are worst affected by irregular water supply from government department. In these areas hand pumps are also not working properly because underground water

Figure 6.20 *Tonk: Quality of Life Based on Drinking Water Facility*

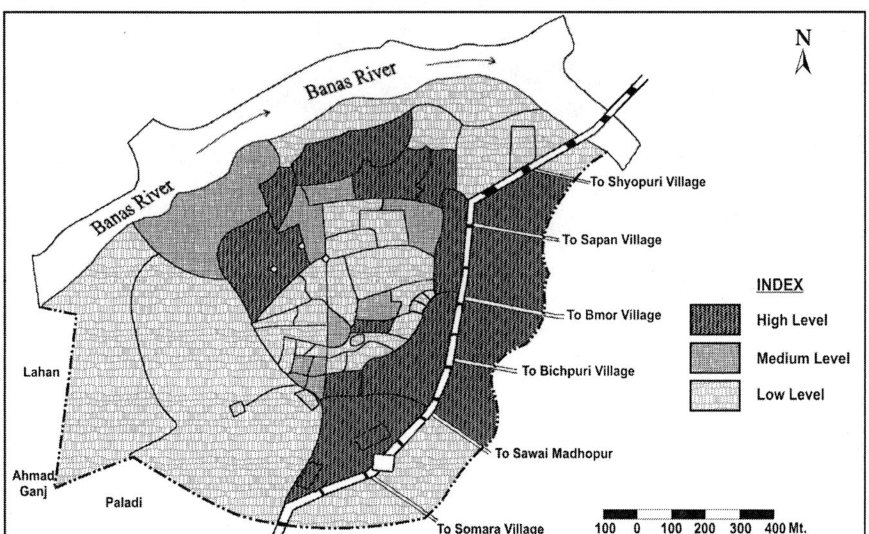

level has gone deeper. The areas of medium level QOL based on drinking water supply are wide spread in the city. But in these areas also water supply is irregular and not sufficient. Only in four wards (ward no. 4, 18, 19 and 21) drinking water facility is good. But on the whole the condition of the availability of drinking water is not satisfactory because of the shortage of water.

QOL Based on Internal Drainage Facility

The condition of internal drainage in Tonk is not proper in comparison to other cities of above one lakh population. This is also because of the fact that in old town streets are very narrow. But a better side of this town is that in many areas slope of the drainage is proper and because of this waste water is not collected in drains for a long time. Our survey reveals that 17 wards of the city are in low category, 24 wards in medium and only 4 wards are in high category. The high QOL is in ward no. 4, 18, 19 and 21. In medium QOL areas drains are Packka but their clearance is not regular and often remained chocked because of the waste disposed by residents (Figure 6.21).

Figure 6.21 *Tonk: Quality of Life Based on Internal Drainage Facility*

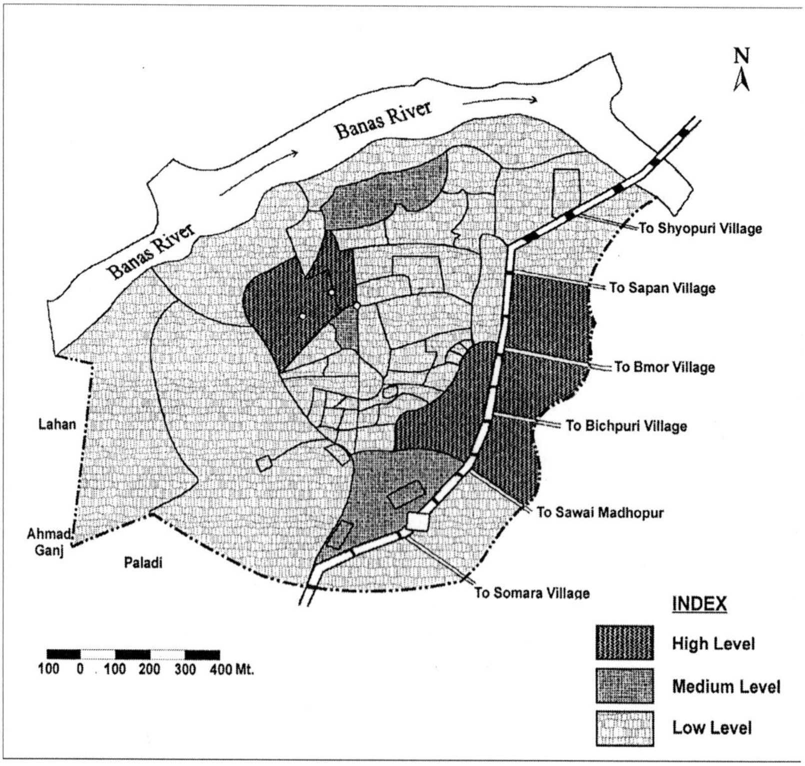

Figure 6.22 *Tonk: Quality of Life Based on Street Conditions*

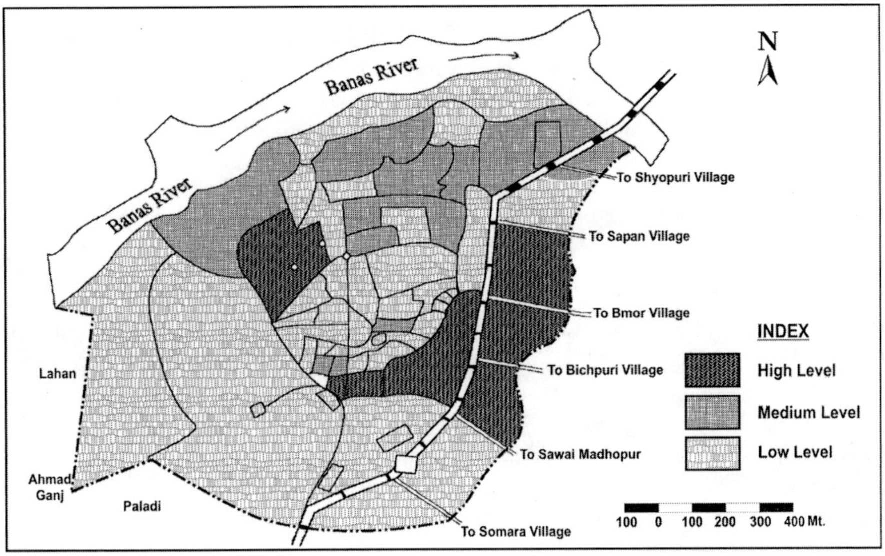

QOL Based on Street Conditions

The spatial pattern of QOL based on street conditions in Tonk follows the same pattern as in case of other variables described above. The high level conditions are in ward no. 4, 18, 19 and 21. These areas are occupied by mostly officers and high income group people. In these areas roads are wide with drains and lights. In medium category as many as 24 wards are included. The road conditions in these areas is satisfactory but in absence of maintenance, nowadays roads have become a cause of accidents. The worst condition of roads are in 18 wards. These areas are widespread in whole of the southern, south-western, northern and central part of the city. In many areas streets are still not paved, what to talk about painted or cement streets.

References

Bederman, S.H. (1974), 'The Stratification of "Quality of Life" in Black Community of Atlanta, Georgia', *Southeastern Geographer*, 14, pp. 26–37.

Berry, B.J.L. (1973), *The Human Consequences of Urbanisation*, Macmillan, London.

Cox, K.R. (1972), *Man, Location and Behaviour: An Introduction to Human Geography*, John Wiley & Sons, New York.

Dickinson, J.C., R.J. Gray and D.M. Smith (1972), 'The "Quality of Life" in Gainesville, Florida: An Application of Territorial Social Indicators', *Southeastern Geographer*, 12, pp. 121–32.

Drewnowski, J. (1970), *Studies in the Measurement of Levels of Living and Welfare*, UN Research Institute for Social Development, Geneva.

Eyles, J. (1974), 'Social Theory and Social Geography', *Progress in Geography*, 6, pp. 27–87.

Howe, G.M. (1972), *Man, Environment and Disease in Britain*, David and Charles, Newton Abbot and Barnes & Nobel, New York.

Knox, P.L. (1975), *Social Well-Being: A Spatial Perspective*, University Press, Oxford.

Koelle, H.H. (1974), 'An Experimental Study on the Determination of a Definition for the "Quality of Life"', *Regional Studies*, 8, pp. 1–10.

Liu, P.C. (1973), *The Quality of Life in the United States 1970: Index, Rating and Statistics*, Midwest Research Institute, Kansas City, MO.

Nath, S.K. (1973), *A Perspective of Welfare Economics*, Macmillan, London.

Ontell, R. (1973), *The Quality of Life in San Diego*, The Urban Observatory of San Diego, San Diego.

Smith, D.M. (1973), *The Geography of Social Well-Being in the United States*, McGraw-Hill, New York.

— (1977), *Human Geography: A Welfare Approach*, Edward Arnold, London.

Terleckyj, N.E. (1975), *Improvements in the Quality of Life and Estimates of Possibilities in United States*, National Planning Association, Washington, D.C.

Wilson, J.O. (1969), *Quality of Life in the United States: An Excursion into the New Frontiers* of *Socio-Economic Indicators*, Midwest Research Institute, Kansas City, MO.

7 Levels of Quality of Life in Urban Centres

In the preceding chapter a detailed analysis of QOL has been done in selected urban centres of Rajasthan. The spatial pattern of QOL has been determined on the basis of five variables, i.e. housing conditions, street conditions, drinking water facilities, internal drainage facilities and waste disposal facilities. The areas of high, medium and low QOL have been delimited and characteristics of each type of area have been discussed on the basis of each variable. But there is a need to delimit the overall or general pattern of QOL in cities. In other words levels of QOL or hierarchy of the QOL needs to be determined on the basis of all the variables, which will not only help in explaining the spatial pattern of urban QOL but also help in urban planning. Once areas of high, medium and low quality of life are determined, planning for their improvement can be done. The objective of the present chapter is to determine the levels of QOL in sample cities and to explain their characteristics.

Methodology

In the last chapter wardwise QOL in terms of housing conditions, street conditions, drinking water facilities, internal drainage facilities and waste disposal facilities has been determined and areas of high, medium and low QOL have also been identified. For each variable having above 75 per cent value has been considered as high, between 50 to 75 per cent as medium and below 50 per cent as low. The same has also been depicted on maps for each variable. The general spatial pattern of the levels of QOL in cities has been determined as follows. For each variable in each ward if the value is high, 5 index value points have been allotted, if its value is medium, the index value will be 3 and for low the index value has been given as 2. Accordingly for all the five variables for each ward index value has been allotted as 5 or 3 or 2. The total of index value points forms the basis of the rank or level of QOL in each ward. The basis followed for wardwise status is as follows:

High QOL	–	Above 20 Index value
Medium QOL	–	15 to 20 Index value
Low QOL	–	Below 15 Index value

Accordingly, level of QOL for all the wards of the city has been tabulated and the same has been depicted on city map. This exercise is simple but effective in determination of the levels of QOL in sample cities of Rajasthan. On the basis of above mentioned methodology, levels of QOL in Kota, Bikaner, Sriganganagar, Bharatpur and Tonk have been determined and discussed in this chapter.

General Spatial Pattern of the QOL or Levels of the QOL in Kota

Kota ranks third among the cities of Rajasthan. According to 2001 census, its total population was 7,04,731. Kota is a historical city of Rajasthan and a fast growing city not only of the Hadoti region but of the State. It represents a combination of traditional and modern city and is now growing as an education-cum-industrial city. Therefore the study of the quality of life in Kota will help in explaining the various facets of the urban QOL. The QOL based on selected variables has already been discussed in earlier chapter. On the basis of those variables and methodology, the levels of QOL (wardwise) have been determined as given in Table 7.1 and also shown in Figure 7.1.

Table 7.1 *Kota: Ranking of Wardwise Pattern of QOL*

Ward No.	*Variables and Ranking based on Index Values*						
	Housing Conditions	*Street Conditions*	*Drinking Water Facilities*	*Drainage Facility*	*Waste Disposal Facility*	*Total Index Value*	*Status of QOL*
1	2	2	2	2	2	10	Low
2	2	2	2	2	2	10	Low
3	2	3	2	3	2	12	Low
4	3	5	3	3	5	19	Medium
5	3	5	5	5	5	23	High
6	3	3	5	2	3	16	Medium
7	2	2	2	2	2	10	Low
8	3	3	3	2	2	13	Low
9	2	3	3	2	2	12	Low
10	2	2	2	2	2	10	Low
11	2	2	2	2	2	10	Low
12	3	5	5	5	5	23	High
13	2	2	2	2	2	10	Low
14	3	5	5	5	5	23	High
15	5	5	3	3	5	21	High
16	3	5	5	5	5	23	High
17	2	2	2	2	2	10	Low
18	2	2	2	2	2	10	Low
19	3	3	3	3	3	15	Medium

Cont'd...

...Cont'd

20	5	3	3	3	3	17	Medium
21	5	3	3	3	5	19	Medium
22	2	5	3	3	5	18	Medium
23	5	3	5	5	5	23	High
24	3	3	3	3	3	15	Medium
25	3	3	3	3	3	15	Medium
26	3	2	2	2	3	12	Low
27	3	3	3	3	3	15	Medium
28	3	3	3	3	3	15	Medium
29	3	3	2	3	3	14	Low
30	3	3	3	3	3	15	Medium
31	2	2	2	2	2	10	Low
32	2	2	2	2	2	10	Low
33	2	2	2	2	2	10	Low
34	5	5	5	5	5	25	High
35	5	5	5	5	5	25	High
36	3	5	2	3	2	15	Medium
37	3	5	3	3	3	17	Medium
38	3	5	5	5	3	21	High
39	5	5	3	3	3	19	Medium
40	5	5	5	5	5	25	High
41	5	5	5	5	3	23	High
42	3	5	5	3	3	19	Medium
43	3	2	3	2	2	12	Low
44	2	2	2	2	2	10	Low
45	2	2	2	2	2	10	Low
46	2	2	2	2	2	10	Low
47	2	2	2	2	2	10	Low
48	2	2	2	2	2	10	Low
49	5	5	5	5	5	25	High
50	5	5	5	5	5	25	High
51	5	5	5	5	5	25	High
52	4	4	4	4	4	20	Medium
53	4	5	5	4	4	22	High
54	5	5	5	5	5	25	High
55	3	5	3	3	3	17	Medium
56	5	5	5	5	5	25	High
57	5	5	5	5	3	23	High
58	3	5	5	3	3	19	Medium
59	5	5	5	5	3	23	High
60	5	5	5	5	3	23	High

Figure 7.1 *Kota: Quality of Life, General Spatial Pattern*

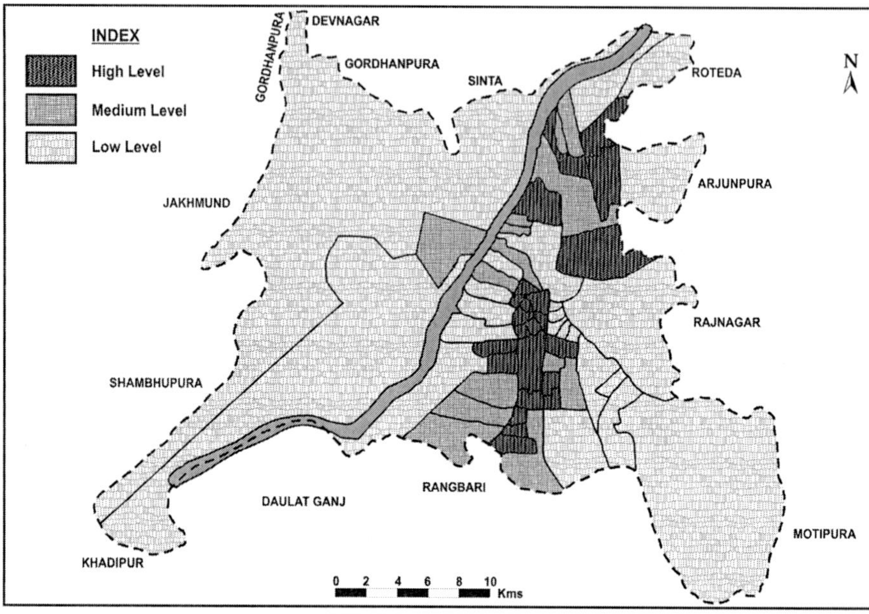

It becomes clear from the table that the spatial pattern of the three categories of QOL in Kota, i.e. high, medium and low is varied as shown in Figure 7.1. The wardwise level of QOL is as follows:

Levels of QOL	*Ward Numbers*
High	5, 12, 14, 15, 16, 21, 23, 34, 35, 38, 40, 41, 49, 50, 51, 52, 53, 54, 56, 59 and 60
Medium	4, 6, 19, 20, 22, 24, 25, 28, 29, 30, 36, 37, 39, 42, 55 and 58
Low	1, 2, 3, 7, 8, 9, 10, 11, 13, 14, 16, 17, 26, 29, 31, 32, 33, 43, 44, 45, 46, 47 and 48

High Level of QOL

In Kota city high level of quality of life is in 21 wards. These wards are widespread in the city and located mostly in Kota north, central and southern parts. In Kota north army area, civil lines, officers colony of railway, MBS Nagar have been included in this category. In central Kota colonies developed outside walled city such as Gumanpura, Vallabhbari, Shopping Centre, New Dhan Mandi have been included. While in new Kota Talwandi, Dadabari, Mahaveer Nagar, Jawahar Nagar, Basant Vihar, R.K. Puram and other new colonies in these areas have been included. In these areas all the houses are well built with proper ventilation facility. The streets are either CC roads or BT roads. The drainage system is also proper and drinking water facilities are also good. In fact, these areas are either occupied by government officers or by high income group of people. In these areas also there are some spots having low or medium QOL, but on the whole these areas are having high level of QOL.

Medium Level of QOL

In Kota areas having medium QOL are spread in 16 wards. In these wards index value varies from 15 to 20. The areas included in this category are Nayapura Bus Stand area (ward 27), Ladpura, Rampura, Khai road area near MBS Hospital, Kishorepura, some areas of Kunahadi, Keshavpura, parts of Indira Vihar, Basant Vihar, Vallabh Nagar, Vigyan Nagar and also southern parts of the city, i.e. ward no. 4, 6, 36, 37 and 42. In these areas although houses are *Packka* but their size is medium, i.e. 1,000 to 1,500 sq. ft. In many areas houses are very close to each other and congested. There is no open space in front of the house, but 90 per cent houses are having regular water supply and electric connections. The street conditions are varied but satisfactory. The waste collection and disposal facility, although available but their periodicity is longer, i.e. twice in a week. The internal drainage is also there but sewerage system is not in function and most of the drains remain choked and become a problem in rainy season.

Low Level of QOL

In Kota low level of QOL has been observed in 23 wards. These areas include whole of the western part of the city located west of river Chambal, most of the old city, i.e. Chanraghata, Tipta, Kishorpura, Ghantaghar, Bajaj Khana, ward no. 13, 17 and 18 of north Kota, Kotadi, Kansuwa, Anantpura, Kachhi Basti located in Indraprastha Industrial area, Shriram Nagar, Prem Nagar, etc. In these areas size of the houses is less than 1,000 sq. ft and 80 per cent houses are of 200 to 250 sq. ft size. In fact, these areas are not having healthy environment and are considered as disease prone areas. The streets are narrow and drains are either absent or closed by one reason or other. In Kota there are 84 slum areas (*Kachhi Basti*) in which about 1,40,000 people are living and having poor QOL. The big slum areas of Kota are Ram Nagar, Ganesh Nagar, Balakund, Durga Basti, Shivpura, Rangbari, Santoshi Nagar, Anantpura, Deendayal Upadhaya Nagar, Near Dakania Railway Station, Bapu Nagar, Gobria Bavadi, Indra Colony, Surya Nagar, Raipura, Thekra, Bajrang Nagar, Adarsh Nagar, Dostpura, Adharshila, Dadabari Oriya Basti, Talwandi Oriya Basti, Sur Sagar, Govind Nagar, along river Chambal, Sazi Dehra, Chhatarpura village, etc. All these areas are having low level of QOL.

Bikaner

The spatial pattern of QOL in Bikaner has been determined on the basis of methodology, mentioned earlier in this chapter. The wardwise pattern of

QOL has been determined as given in Table 7.2 and the same has been depicted in Figure 7.2.

Table 7.2 *Bikaner: Ranking of Wardwise Pattern of QOL*

Ward No.	Housing Conditions	Street Conditions	Drinking Water Facilities	Drainage Facilities	Waste Disposal Facilities	Total Index Value	Status of QOL
1	2	2	2	2	2	10	Low
2	3	3	5	3	3	17	Medium
3	2	2	5	2	2	13	Low
4	5	5	5	3	3	21	High
5	5	5	5	5	5	25	High
6	2	5	3	3	3	16	Medium
7	5	5	5	5	5	25	High
8	5	5	5	2	2	19	Medium
9	5	5	5	5	5	25	High
10	5	5	5	5	5	25	High
11	5	5	5	5	5	25	High
12	5	5	5	5	5	25	High
13	2	2	2	2	2	10	Low
14	5	2	2	2	2	13	Low
15	3	2	5	3	3	16	Medium
16	3	3	5	3	3	17	Medium
17	2	2	5	3	3	15	Medium
18	2	5	5	3	3	18	Medium
19	5	5	5	5	5	25	High
20	5	5	5	5	5	25	High
21	2	2	5	2	2	13	Low
22	2	2	3	3	2	12	Low
23	5	5	5	5	5	25	High
24	5	5	5	5	5	25	High
25	2	5	3	3	3	16	Medium
26	2	2	2	2	2	10	Low
27	2	2	2	2	2	10	Low
28	3	3	2	2	2	12	Low
29	5	5	5	5	5	25	High
30	3	2	3	2	2	12	Low
31	3	3	3	3	3	15	Medium
32	3	3	3	3	3	15	Medium
33	3	3	3	3	3	15	Medium
34	3	3	3	3	3	15	Medium
35	3	3	3	3	3	15	Medium
36	5	2	3	3	3	16	Medium

Cont'd...

...Cont'd

37	5	5	3	3	3	19	Medium
38	5	5	5	3	3	21	High
39	5	5	5	2	5	22	High
40	5	5	3	3	5	21	High
41	2	2	3	2	2	11	Low
42	3	2	2	2	2	11	Low
43	3	2	2	2	2	11	Low
44	2	2	2	2	2	10	Low
45	3	3	3	3	3	15	Medium
46	5	3	3	3	3	17	Medium
47	5	3	3	3	3	17	Medium
48	5	3	3	5	5	21	High
49	5	3	3	5	5	21	High
50	5	5	3	5	5	23	High
51	5	5	5	3	3	21	High
52	3	2	2	2	2	11	Low
53	2	2	2	2	2	10	Low
54	5	5	3	3	3	19	Medium
55	5	5	3	3	3	19	Medium

On the basis of above table, the pattern emerged in Figure 7.2 shows the spatial variation of the levels of QOL in Bikaner. The pattern of quality of life in Bikaner and their characteristics is as follows:

Figure 7.2 *Bikaner: Quality of Life, General Spatial Pattern*

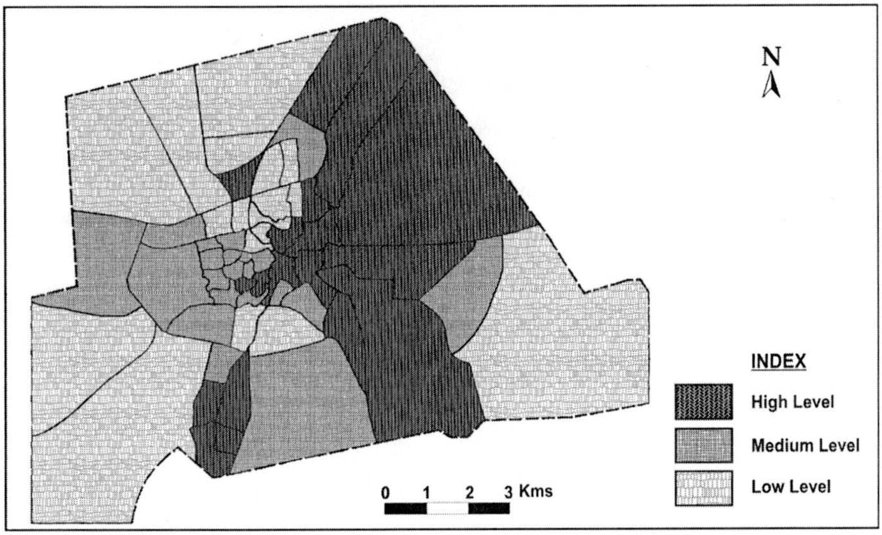

High Level of QOL

These areas are better and differ from the rest of the areas in terms of housing conditions and infrastructural facilities. Congestion and crowding are relatively less and exclusive housing facilities are not uncommon. Owner occupancy is common and room density is less. In these areas houses are well built and provided with amenities such as water supply, electricity, modern baths and latrines, gap, etc. More than 60 per cent houses are having open space in and/or outside the house. These areas are J.N. Vyas Colony, Military Area, Sadulganj, Naganeeliji, Pawan Puri Colony, UIT Colony, Civil Lines, Indira Chowk, Gangashahar, R.C.P. Colony, New Railway Colony, etc.

Medium Level of QOL

The areas having medium level form a transitional zone between low and high class areas. The residents of these areas belong to variety of occupational and cultural groups. A majority of them are engaged in services. Many of them are small shop owners and engaged in petty trade and business. These are mixed residential areas where migrants, generally in government services, live together with the house owners. As compared to the low class areas, the livability in these areas is of a moderate order. Houses in this area are generally brick houses and many of them are quite old. Housing congestion and over crowding too is beyond permissible limits. Exclusive housing facilities are lacking here. However, the level of infrastructural facilities and other amenities can be put as moderate. In Bikaner such areas are Laxmi Nath ji ki Ghati, Bhatron ka Chowk. Acharyon ka Chowk, Benisar well, outside Usten ki bari, inside Idgah bari, etc.

Low Level of QOL

These areas exhibit slum conditions. Population is generally engaged in low paid services. Housing congestion, old dilapidated buildings, lack of open spaces, narrow and winding lanes are physical characteristics of this part. Drainage is poor and rain water pools here during monsoon and breeds mosquitoes. General sanitary conditions are also substandard. Renter occupancy, sharing of living space and lack of housing facilities spread in these parts of the city. Infrastructural facilities and amenities are lacking. In these areas 80 per cent households are not having direct water supply in the house and other amenities are also lacking here. The living condition in these areas is very poor. Most of these areas are the slums of the city. The areas included in this category are Bangalangar, Pabu Chowk. Harijan Basti, Gangashahar, Sangalpura, Rampura Harijan Basti, Goga gate Harijan Basti, Amarsinghpura, Bhuton ka Bas, Subhashpura, Pratap Basti and Sarvodaya Basti (Narsingh Sagar Talab), etc.

Sriganganagar

Sriganganagar is a newly developed town of Rajasthan. It was established in 1927 AD and now has grown as an important trade centre of the State. Srigangagnagar also represents a town developed after the development of irrigation, thus an example for the other towns which are developing in canal command area of Indira Gandhi Canal. The town is a planned town, but like any other town of the State have/had varied levels of QOL in terms of variables discussed earlier. The QOL in Sriganganagar according to selected variables has already been discussed in earlier chapter. The general spatial pattern of the city based on composite index values has been given in Table 7.3 and also depicted in Figure 7.3.

Table 7.3 *Sriganganagar: Ranking of Wardwise Pattern of QOL*

Ward No.	*Variables and Ranking based on Index Values*						
	Housing Conditions	*Street Conditions*	*Drinking Water Facilities*	*Drainage Facility*	*Waste Disposal Facility*	*Total Index Value*	*Status of QOL*
1	5	5	5	5	3	23	High
2	5	5	5	5	5	25	High
3	2	2	2	2	2	10	Low
4	5	5	3	3	3	19	Medium
5	5	5	3	3	3	19	Medium
6	5	5	3	3	3	19	Medium
7	5	5	3	3	3	19	Medium
8	3	5	5	3	3	19	Medium
9	3	2	2	2	2	11	Low
10	2	2	2	2	2	10	Low
11	2	3	2	2	2	11	Low
12	3	2	2	2	2	11	Low
13	5	3	3	2	3	16	Medium
14	3	3	3	3	3	15	Medium
15	5	3	3	3	3	17	Medium
16	5	5	5	5	3	23	High
17	5	5	5	3	3	21	High
18	5	3	3	3	3	17	Medium
19	3	5	3	3	3	17	Medium
20	5	3	3	3	3	17	Medium
21	3	3	5	5	3	19	Medium
22	5	5	3	3	3	19	Medium
23	5	3	3	3	3	17	Medium
24	3	3	2	2	2	12	Low

Cont'd...

...Cont'd

25	5	3	3	3	3	17	Medium
26	3	3	3	3	3	15	Medium
27	3	3	3	3	3	15	Medium
28	3	2	2	2	2	11	Low
29	2	2	2	2	2	10	Low
30	2	2	2	2	2	10	Low
31	3	2	2	2	2	11	Low
32	5	3	3	3	3	17	Medium
33	3	3	3	3	3	15	Medium
34	5	3	3	3	3	17	Medium
35	3	2	2	2	2	11	Low
36	3	2	2	2	2	11	Low
37	3	2	2	2	2	11	Low
38	2	2	2	2	2	10	Low
39	2	2	2	2	2	10	Low
40	2	2	2	2	2	10	Low

The wardwise pattern of the level of QOL becomes clear from the table. The pattern emerged is as follows:

Levels of QOL	*Ward Numbers*
High	1, 2, 16 and 17
Medium	4, 5, 6, 7, 8, 13, 14, 15, 18, 19, 20, 21, 22, 23, 25, 26, 27, 32, 33 and 34
Low	3, 9, 10, 11, 12, 24, 28, 29, 30, 31, 35, 36, 37, 38, 39 and 40

High Level of QOL

In Sriganganagar high level of QOL exists only in four wards, i.e. ward no. 1, 2, 16 and 17. The areas of high level of QOL include Civil Lines, Sukhadia Circle, Jawahar Nagar and along Hanumangarh road. In Civil Line area most of the houses are government quarters. In these areas high level officers are residing. Since these houses were built on British tradition, they have enough open space around the house and also having all modern facilities required for high QOL in terms of selected variables. This area is a green area of the city having too much trees, thus the overall environment of this area is healthy. Another area of high QOL is the Jawahar Nagar area, a planned area developed only after 1985. In this area the built of houses is good and nearly 80 per cent houses having open space in front of them. The street conditions are also good and drainage facilities and waste collection and disposal are also available. Most of the residents of this area are either businessmen, doctors, engineers, and other higher service class people.

Figure 7.3 *Sriganganagar: Quality of Life, General Spatial Pattern*

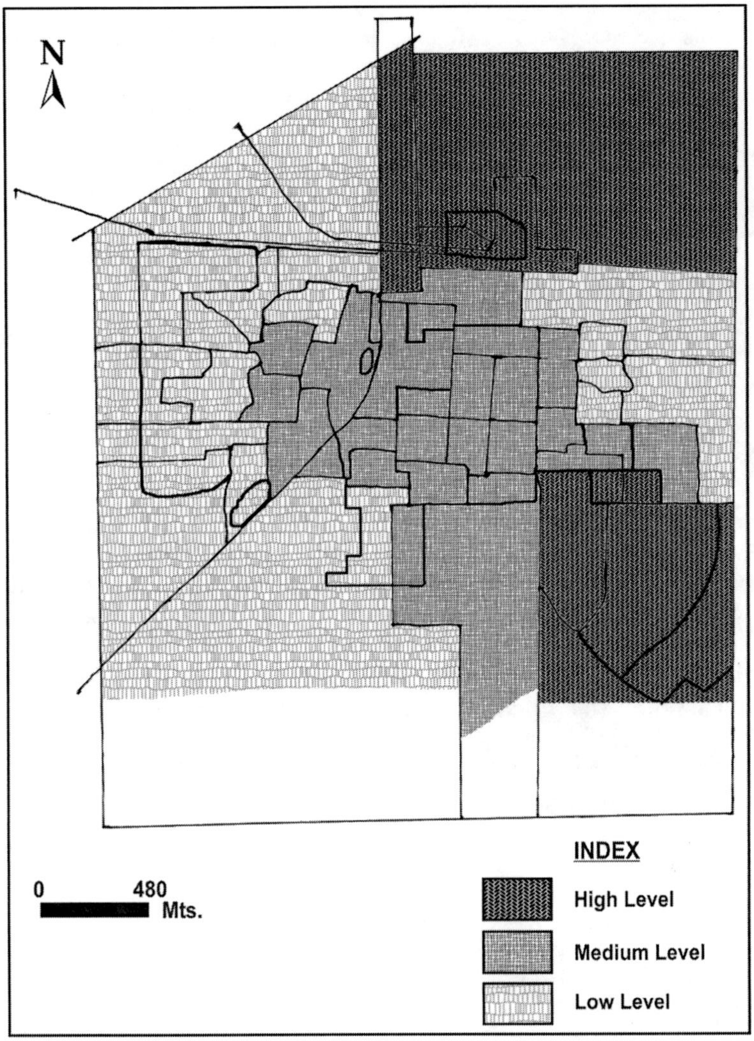

Medium Level of QOL

In Sriganganagar as many as 20 wards are having medium level of QOL, in other words 50 per cent of the city is under this category. This is because of the fact that most of the city population belongs to the middle class. The areas included in this category are entire Block Area (the city is having several blocks such as B, H, F, L, N, etc.). The entire block area is a planned area, but the size of the house in this area is mostly 50'×50' and nowadays reduced to 25'×50'. There is no open space between the houses as well as in front of the house. Otherwise the area is quite good in terms of street conditions, drinking water facilities, drainage and waste collection facilities. The other areas included in this category are ward no. 26, 32, 33 and 34 of Purani Abadi, localities along Padampur, Suratgarh and Hanumangarh roads, Vinoba Basti, Satia Basti, Prem Nagar, Khalsa Nagar, Kunj Vihar, etc. In these areas houses are having proper facilities and healthy atmosphere but conditions of internal drainage and waste collection are not satisfactory.

Low Level of QOL

Although Sriganganagar is a new town but as many as 16 wards are in the category of low level of QOL. The areas of low quality of life are widespread in Purani Abadi. Purani Abadi is that part of the town which was developed before the planned town. In due course of time several colonies (*Basties*) have developed in this part. All these colonies are unplanned. The streets in these areas are very narrow and nearly 30 to 40 per cent houses are still *Kachha*. The size of the houses varies form 10'×10' to 30'×40'. There is no drainage facilities in about 60 per cent of the area.

The collection of waste is also a casual feature. The areas developed along minor canal, along Kanpur road, near JCT Mill, Dev Nagar, Arjun Nagar, Bharat Nagar, Labour Colony, Subhash Nagar, Kedar Nagar, Shyam Nagar, etc. are having low QOL. Another area of low QOL is located in ward no. 3, 9, 10, 11 and 12 and also in southern part, i.e. ward no. 24. The area near Sugar Mill is a big slum area of the city. In this area the living conditions are much below the low QOL. The houses are muddy and their size is hardly 8'×10'. There is no ventilation facility, no drains, no drinking water facility in the house and the whole atmosphere of this area is unhealthy and polluted. Adjacent to this area also low QOL is prevalent but having little more facilities. In fact, the areas mentioned above are the slum areas of Sriganganagar and also disease prone areas.

Bharatpur

Bharatpur is located in the Eastern plain of Rajasthan and is a progressive city of Rajasthan. The city is a combination of old and traditional type of residential areas and newly developed colonies. With the growth of city and infrastructural development some areas have developed properly, while others remained underdeveloped. The present analysis reveals the general QOL in Bharatpur and its spatial pattern.

The levels of QOL in Bharatpur based on selected variables are shown in Table 7.4. On the basis of the table, the areas are included in high, medium and low category and their areas are as follows:

Levels of QOL	*Ward Numbers*
High	5, 12, 18, 21, 23, 24, 29, 36, 40 and 45
Medium	4, 9, 10, 11, 14, 15, 16, 19, 22, 25, 26, 31, 32, 33, 35, 37, 38 and 44
Low	1, 2, 3, 6, 7, 8, 13, 17, 20, 27, 28, 30, 34, 39, 41, 42 and 43

The general spatial pattern of the levels of QOL in Bharatpur have been depicted in Figure 7.4.

Table 7.4 *Bharatpur: Ranking of Wardwise Pattern of QOL*

Ward No.	*Variables and Ranking based on Index Values*						
	Housing Conditions	*Street Conditions*	*Drinking Water Facilities*	*Drainage Facility*	*Waste Disposal Facility*	*Total Index Value*	*Status of QOL*
1	2	2	2	2	2	10	Low
2	2	3	2	2	2	11	Low
3	3	3	3	2	3	14	Low
4	3	3	3	3	3	15	Medium
5	5	5	5	5	5	25	High
6	2	2	2	3	2	11	Low
7	2	2	2	2	2	10	Low
8	2	3	2	3	3	13	Low
9	3	5	3	2	3	16	Medium
10	3	5	3	3	3	17	Medium
11	3	5	3	3	3	17	Medium
12	3	3	5	5	5	21	High
13	3	2	3	2	3	13	Low
14	3	3	5	3	5	19	Medium
15	3	3	3	5	3	17	Medium
16	3	5	3	3	3	17	Medium
17	2	3	2	5	2	14	Low
18	3	3	5	5	5	21	High
19	3	3	3	5	3	17	Medium
20	2	2	3	2	2	11	Low
21	5	5	5	5	5	25	High
22	3	3	3	3	5	17	Medium
23	5	5	5	5	5	25	High
24	5	5	5	3	5	23	High
25	3	3	3	3	5	17	Medium
26	3	5	3	3	3	17	Medium
27	2	2	2	2	2	10	Low
28	2	2	2	2	2	10	Low
29	5	5	5	2	5	22	High
30	2	3	2	3	3	13	Low
31	2	3	2	5	2	14	Low
32	2	3	2	5	2	14	Low

Cont'd...

...Cont'd

33	3	5	3	3	3	17	Medium
34	2	3	2	2	2	11	Low
35	3	3	3	3	3	15	Medium
36	5	5	5	2	5	22	High
37	3	3	3	3	3	15	Medium
38	3	3	3	3	5	17	Medium
39	2	2	2	2	2	10	Low
40	5	5	5	3	5	23	High
41	2	3	2	3	2	12	Low
42	2	3	2	2	2	11	Low
43	2	3	2	2	2	11	Low
44	3	5	3	3	3	17	Medium
45	5	5	5	5	5	25	High

Figure 7.4 *Bharatpur: Quality of Life, General Spatial Pattern*

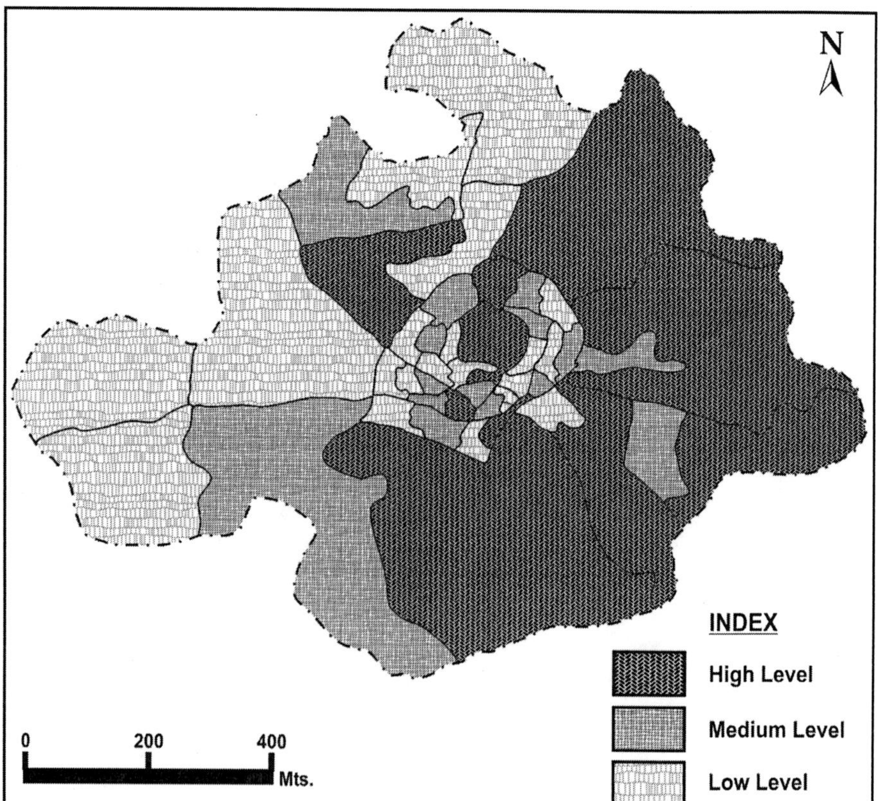

High Level of QOL

In Bharatpur city 10 wards are having high level of QOL. These areas are mostly located in eastern and south-eastern part of the city. A small area in central part is also located. These areas are mostly located away from the core area of the city and developed as new colonies after 1985. In these areas officers of various departments and businessmen are residing. Their houses are well built and 80 per cent houses in these areas are having area more than 2,000 sq. ft. Nowadays ward no. 21 is considered as the best locality of the city. In these areas not only built of the house is good but having open space and all other facilities required for good QOL. The water supply is regular, similar is the case of waste disposal facility and internal drainage facility. The streets in these areas are either cemented or metalled. In brief, these areas are environmentally healthy.

Medium Level of QOL

In Bharatpur 18 wards are having medium QOL in Bharatpur. Most of these wards are located in central part of the city and only two wards are located in south-western and north-western part. The central part of the city is the old part of the city and having traditional culture. The size of the house in this part varies from 1,000 to 2,000 sq. ft and nearly 80 per cent houses are of double or triple storey. The width of the streets are also not proper. The general width is between 8 to 10 feet, but certain areas are even having streets of lesser width. Although, most of streets are either cemented or metalled. The internal drainage system is there but their clearance is not regular. Most of the drains remain chocked with domestic waste, for this people residing there are also responsible. The drinking water facility is good but waste collection and disposal facility is not proper, therefore, heaps of domestic and municipal waste can be seen here and there. Some of the colonies developed outside the central area of the city are good but in some cases drinking water supply is not regular. Most of these colonies have been occupied by middle class government servants, businessmen and other service class people.

Low Level of QOL

In Bharatpur as many as 17 wards are having low QOL. These areas are located in central as well as eastern and another part of the city. These areas are the '*Kachhi Basti*' or slum area of the city. The areas included in this category are Harijan Basti (Anah Gate), Dhimier – Kalwar Basti (Sona Ram Gate), Mori Mohalla (Char Bag), Jatav Basti, Harijan Basti (Papadi Mohalla), Harijan Basti (Suraj Pole), Jatav Basti (Suraj Pole), Jatav Basti (Jadhina), Mali Mohalla, Kachha Kunda, Harizan Basti (Mathura Gate

Kumhor Gate and Gordhan Gate) and Nagla Toojani. In these areas general conditions are poor. The housing conditions are not proper, the size of the houses is less than 25 sq. ft. Most of the houses are only single room houses having no latrines or bathroom. There is no water connection in the houses. The street conditions are also poor. The streets are in the form of lanes with a width of about 4 to 5 feet. The internal drainage is either absent or choked with waste material. During rainy season the streets become drains of rain and dirty water. The overall environment in these areas is not healthy and the areas are highly disease prone areas.

Tonk

Among above one lakh population cities of Rajasthan, Tonk is having a population of 1,35,689 in 2001. Tonk is still a traditional type of city. Although, National Highway No. 12 passes through the city but in absence of railways its development remained stagnant. It was only after 1980 the city has started to expand its limit in southern and eastern side. But in spite of development still Tonk is considered as a traditional city of the State, therefore, it is interesting to study the QOL. The QOL in Tonk based on each selected variable has already been discussed in detail in last chapter. The general levels of the QOL or spatial pattern of the levels of QOL in Tonk have been tabulated in Table 7.5.

Table 7.5 *Tonk: Ranking of Wardwise Pattern of QOL*

Ward No.	*Variables and Ranking based on Index Values*						
	Housing Conditions	*Street Conditions*	*Drinking Water Facilities*	*Drainage Facility*	*Waste Disposal Facility*	*Total Index Value*	*Status of QOL*
1	3	3	3	3	3	15	Medium
2	2	2	2	2	2	10	Low
3	2	2	2	2	2	10	Low
4	3	2	2	2	2	11	Low
5	3	3	5	2	2	15	Medium
6	3	3	3	3	3	15	Medium
7	5	3	2	2	3	15	Medium
8	3	2	2	2	2	11	Low
9	2	2	2	2	2	10	Low
10	2	2	2	2	2	10	Low
11	2	2	2	2	2	10	Low
12	3	3	3	3	3	15	Medium
13	3	3	3	3	3	15	Medium
14	5	3	3	3	3	17	Medium
15	3	3	3	3	3	15	Medium

Cont'd...

...Cont'd

16	3	3	3	3	3	15	Medium
17	2	2	2	2	2	10	Low
18	5	5	4	3	5	22	High
19	5	5	5	3	3	21	High
20	3	3	3	3	3	15	Medium
21	5	5	5	3	3	21	High
22	5	3	3	2	2	15	Medium
23	3	5	5	3	2	18	Medium
24	2	2	2	2	2	10	Low
25	2	2	2	2	2	10	Low
26	2	2	2	2	2	10	Low
27	2	2	2	2	2	10	Low
28	5	3	3	3	2	16	Medium
29	3	3	3	3	3	15	Medium
30	3	3	3	3	3	15	Medium
31	5	3	3	2	2	15	Medium
32	2	2	2	2	2	10	Low
33	2	2	2	2	2	10	Low
34	5	3	3	2	2	15	Medium
35	5	3	3	2	2	15	Medium
36	5	3	3	2	2	15	Medium
37	3	3	2	2	2	12	Low
38	2	2	2	2	2	10	Low
39	5	3	3	2	2	15	Medium
40	5	3	3	2	2	15	Medium
41	5	3	3	3	2	16	Medium
42	3	5	5	2	2	17	Medium
43	5	3	5	2	3	18	Medium
44	2	2	2	2	2	10	Low
45	2	2	2	2	2	10	Low

The pattern emerged from the above table is as follows:

Levels of QOL	*Ward Numbers*
High	18, 19 and 21
Medium	1, 5, 6, 7, 12, 13, 14, 15, 16, 20, 22, 23, 28, 29, 30, 31, 34, 35, 36, 39, 40, 41, 42 and 43
Low	2, 3, 4, 8, 9, 10, 11, 17, 24, 25, 26, 27, 32, 33, 37, 38, 44 and 45

The above levels of QOL in Tonk have also been depicted in Figure 7.5.

Figure 7.5 *Tonk: Quality of Life, General Spatial Pattern*

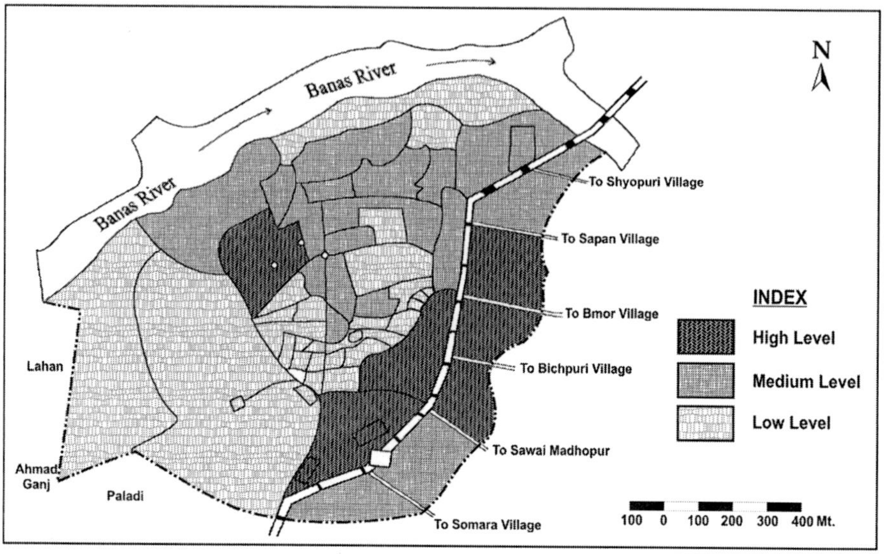

High Level of QOL

In Tonk only three wards are having high QOL. The areas included in these wards are Civil Lines, Adarsh Nagar, Indra Colony, Housing Board Colony and Shastri Nagar. The built of the houses in this area is good and space between houses has been provided. Although a few houses have open space in front of them. But houses in this area have other modern facilities. The size of the house varies from 1,800 sq. ft to 3,200 sq. ft. The road conditions are good and internal drainage system is also good. There is a general slope in this area, therefore, water in drains easily flow down and drains remain clean in most of the cases. The water supply is also good in comparison to other parts of the city. But there is a general scarcity of water in the city, therefore, these areas also receive limited supply. In these areas most of the people residing are government and other officers, businessmen and other families of higher income group.

Medium Level of QOL

Out of 45 wards in the city as many as 24 wards are having medium QOL. The main areas included in this category are Mahadewali, Bahir, Santosh Nagar, Takhta, Shanti Nagar, Chawani, Bambor gate, Sindhi Colony, Mehandi Bagh, Phool Bhag, Childdi Ki Badi, Bada Kuia, Panch Batti, Begam Bag, Ahuja Colony, Chudigaran, Santosh Nagar, etc. The size of houses in these areas varies from 500 to 2,000 sq. ft, although all the houses are *Packka* in nature but not having open space and congested. The streets are also not much wide. The internal drainage exists but in absence of clearance most of the drains remain chocked. Similarly the system of waste collection and disposal is not proper and waste collection is done on weekly basis. The waste is often dumped in Idgah Talab, Dhanna Talai and Chaturbhuj Talab, which becomes a cause of environmental degradation.

Low Level of QOL

In Tonk 17 wards are having low QOL. Most of these areas are located in central part of the city as well as in south-western (ward 2, 3 and 4) and northern part (ward 44 and 45). The areas included in this category are Hazari Pada, Gandi ki Choki, Kampu Ginsi, Dhanna Talai, Goldungari, Hira Chowk, Harijan Basti, Kalipaltan, Regran Basti, Moti Bagh, Basti Julekhan. These areas are very congested. The houses are not only congested but not having proper ventilation facilities. The houses are quite old and having very small rooms. The size of the family living in these houses is also big. The streets are narrow, at many places only 4 to 5 feet wide. The drains are there but most of them are not in working condition and drain water often spreads on streets. There is no system of waste collection and disposal. In fact, these areas are the *Kachhi Basti* or slum areas of the city. These areas are environmentally not healthy and disease prone areas of Tonk.

8 Problems and Planning

The various facets of urbanisation, degradation of environment and quality of life in cities of Rajasthan have been discussed in last five chapters. The urban population is growing at a fast rate and similarly urban settlements are also increasing in size and number. Cities are important to national and/or regional economies because they are the focal points for important activities like trade, commerce, industry and government administration. Cities are also centres of education, health care, technological innovation, entrepreneurship and governance. But despite these advantages, most of the cities face significant problems. A large segment of urban population suffer from the problems of degradation of environment and lower quality of life. Urban areas in India, including that of the Rajasthan State are characterised by high concentration of economic activities, high density of population, unabated immigration, mushrooming of squatter colonies and increasing scarcity of basic amenities and infrastructures. The urban environment is degrading at a fast rate due to rapid growth of population and cities are facing many problems related to degradation of environment and quality of life. The present chapter deals with the problems of urbanisation, degradation of environment and quality of life in cities of Rajasthan having above one lakh population. In this chapter not only problems have been discussed but suggestions for the improvement of existing problems have also been given and a long term strategy has also been suggested. The identification of problems has been done in sample towns and based on fieldwork. The samples selected and number of respondents in each sample city are as follows:

S. No.	*Sample City*	*Number of Respondents*
1.	Kota	2,600
2.	Bikaner	1,850
3.	Sriganganagar	1,280
4.	Bharatpur	1,200
5.	Tonk	840

There are also certain general problems for which information during fieldwork have been collected by the authors. The problems related to 'Degradation of Environment' especially air, water, land and noise pollution in sample cities have already been discussed in details in Chapter–4. In this section an attempt has been made to explain the growing problems of urbanisation and quality of life.

Problems and Issues

In Rajasthan, cities having one lakh and above population are growing at a fast rate and gradually becoming unmanageable and alarming with the growing inequalities arising out of logs in adjustment to rapid urbanisation. A great many 'urban' dwellers and individual households are not integrated socially, economically or politically in urban life. 'Their low level of incomes, lack of education, sub-standard living in slums and squatter settlements and increasing population concentrations, overloading the community's environmental life support system, in the face of already existing backlog of services and amenities, have added new dimensions to the problem increasing the maintenance of urban development' (Wishwakarma, 1988: 174).

The most important problems of present day urbanisation are:

- Informal growth of urban poor and city's environs, slums and squatter settlements;
- Public health and sanitation;
- Inadequate housing and land control measures;
- Extremely inequitable access to resource including land and capital;
- Land allocation and land use for housing and other sectors of urban uses;
- Inadequacy of parking facilities;
- Problem of drinking water;
- Lack of sewerage and drainage;
- Improper solid waste disposal and management;
- Spatial inequalities in quality of urban life;
- Control of pollution and epidemic diseases;
- Unemployment and underemployment;
- Lack of coordination between city authorities regarding maintenance and development of the city;
- Conservation of basic resources;
- Unplanned growth of cities; and
- Maintenance of law and order and security of the people.

Some of the specific problems emerged from present study are as follows:

Housing

Housing is a basic necessity of human beings. Adequate shelter for each and every household is a fundamental prerequisite to healthy living and quality of life. The Indian Planning Commission has realised that the role of housing in economic development both as an end and as a means, is an important one; as an end, housing is an essential ingredient in the basic requirement of civilised living; as a means, besides adding to national income, it is a strong motivator of savings and plays an important part in the generation of employment with population increasing in the cities, there is a chronic shortage of housing. Since the majority of the population moving into cities belongs to the economically deprived strata, and since there is no provision to accommodate them, this population is compelled to live in improvised shacks, resulting in the growth of slums. These unauthorised hutments usually spring up in the most uninhabitable sites in cities, such as along river channels or *'nallahs'* or on unused land that cannot be put to any productive use. They also occupy open space, thereby nullifying the basic function of such open ground. Most of these colonies, devoid of water supply, drainage and other basic amenities, are veritable cesspools of disease. Shortage of houses is a problem that affects all strata of population in cities. The haphazard and unrestricted growth of housing colonies in the peripheral areas of cities results in not only frequent expansion of city limit, but also the disappearance of green belt. Owing to the pressure of population on land, housing design and layout have dispensed with open spaces around houses, converting the landscape into a mass of concrete. Most of the new housing colonies spring up even before roads, electricity, drinking water supply, drainage and other infrastructure can develop. This leads to illegal tapping of existing resources from the nearest available colonies. Limitation of space is the main physical condition resulting from overcrowding.

Our study reveals that in cities of Rajasthan housing is the main problem. The bigger the city more is the housing problem. Along with housing, the main problem is the substandard housing facilities. The study also reveals that only areas of high QOL are having good housing facilities. This is because of the fact that houses in these areas belong to either higher income group of people or these houses are government houses of officers. The condition in areas of medium level of QOL is also satisfactory. But in low level QOL areas, the housing conditions are worst. Nearly 75 per cent houses in such areas are *Kachha* or mud houses or house built by pilling stones or cottage and in many cases a polythene sheet supported by bamboos. Most of these houses are single room houses in which a family of 6 to 8 persons used to live in an unhygienic environment. In all the cities of Rajasthan such areas can be seen and in cities having above one lakh population about one fourth of the city area is occupied by them. These areas are known as *'Kachhi Basti'* (slum).

Urban Slums

Ashish Bose (1955: 19) defined slum as 'deprived human settlement, a settlement which is demographically, economically vulnerable'. Extreme overcrowding, high density and high levels of mortality and fertility are typical demographic features of slums. A large unorganised sector, low levels of productivity and extreme poverty are the usual economic features, and the lack of access to basic needs like water, sanitation and clean environment make these areas environmentally hazardous.

Slums are common in all the urban areas of Rajasthan. The common characteristics of slum residents are:

(a) The families living in slum and squatter settlements are mostly poor class people;
(b) Most of the squatting people are landless labourers;
(c) Slum dwellers mostly live in areas of low land value due to their low income;
(d) People living in slums are biased with profession in most cases;
(e) People living in squatter areas sometimes suffer from serious diseases. They mostly indulge in unsocial behaviour, and other related problems; and
(f) The occupational structure of slum dwellers includes labours, municipal workers, rickshaw pullers, baggers, waste product collectors, etc.

Our study reveals that in all the sample towns, i.e. Kota, Bikaner, Sriganganagar, Bharatpur and Tonk slums are the biggest problem. Because of the low quality of life, these areas are disease prone areas and also responsible for the degradation of environment. As stated earlier, bigger the size of the city, more are the slum areas. For example, in Kota city 41 areas have been identified as '*Kachhi Basti*'. These are Ram Nagar, Geneshpal, Balakund, Durga Basti, Durga Basti Ext., Shivpura, Rangbari, Santoshi Nagar, Anantpura, Pt. Deendayal Up. Nagar Dakania Rly. Station to Vigyan Nagar, Bapu Nagar, Gobaria Bavdi, Indra Colony A&B, Surya Nagar, around DCM Circle, Raipura, Tipta, Bajrang Nagar, Sanjay Nagar A&B, Dostpura A&B, Brijraj Nagar, Adhershila, Adarsh Nagar, Shyam Nagar, Dadabari Oriya Basti, Keshavpura A&B, Talwandi Oriya Basti, Prem Nagar, Sur Sagar, Ghodewala Baba, along river Chambal, Sazi Dehra and Chhatarpura village.

In fact, slums are common in all the cities having above one lakh population in Rajasthan and having low quality of urban life.

Urban Sanitation

In urban areas of Rajasthan, the condition of sanitation is not satisfactory. This is a common problem of urban India, as Planning Commission reports indicate that at least 52 per cent of the urban population has no access to proper sanitation and about 15 per cent are yet to be provided with safe drinking water. Absence of continuous flow of water enhances the unhygienic conditions.

Absence of a proper sewerage system in the newly developed housing colonies, is one of the major problems of sanitation. It not only leads to deterioration of the quality of areas, but is also a root cause of numerous infectious and parasitic diseases. Lack of paved roads in many of these residential colonies leads to accumulation of water and slush during the rainy season, providing ideal conditions for the breeding of mosquitoes. The puddles of water and heaps of garbage along the roads facilitates the proliferation of disease including organism, such as bacteria and protozoa, vectors including mosquitoes, houseflies, various arthropods and helminthes. The drainage channels in all the cities are usually clogged with wastes and industrial effluents, and reduced to sluggishly flowing cesspools, which pose immense health hazards. Stray animals, including cattle, pigs and dogs, which form a part and parcel of cities, are another source of insanitary conditions, resulting in further health risks.

Drainage and Sewerage System

In all the cities of Rajasthan internal drainage and sewerage system is not satisfactory. The QOL in terms of internal drainage system has been described in Chapter–6 and their spatial pattern has been depicted in Figure 6.3, 6.7, 6.9, 6.12, and 6.21. The analysis reveals that about 20 to 30 per cent areas of the cities are not having internal drainage system. Even where drainage system is there, the clearance is not regular, therefore, the system is not effective and creates problems. In all the sample cities sewerage system is not at all effective. In Kota, Bikaner and Sriganganagar sewerage system has been developed partially, that too not functioning properly. In absence of sewerage plant of proper capacity all the cities are facing great problem.

Apart from environmental problems, the drainage and sewerage situation causes serious problems, which pose the threat of Malaria and other diseases. While open drains have posed a big threat to most of the urban dwellings, the entire absence of a drainage system itself has aggravated the situation. The expanding slums and growing emigration are mostly the cause and effect of deteriorating drainage conditions leading to serious health hazards.

Problems of Waste Collection and Disposal

The main problem of rapid urbanisation is the management of urban wastes, both solid and liquid, which are the biggest potential source of pollution and health hazard. Urban wastes are discarded organic or inorganic substance in the form of solid, semi-solid, liquid and gases which are residues or derivates of human, vegetable material and industry. These wastes include household wastes, municipal wastes, commercial wastes, industrial wastes, liquid wastes and gaseous wastes. Among these, solid waste is piling up and its volume is increasing day by day in all the cities.

According to an estimate the urban refuse amounts to about 0.3 kg of *Kachra* (household-refuse, city wastes and commercial wastes) and 0.2 kg of night soil and 0.8 kg of urine per head per day. There is however considerable variation from centre to centre, depending upon local factors.

In all the towns waste disposal system is grossly inadequate, as a consequence of which huge dumps of garbage on the roadsides, even in upper class areas, are a regular sight in almost all cities. The spatial pattern of waste disposal facilities and the nature of QOL has been discussed in Chapter–6. In all the selected towns waste collection and disposal facility is unsatisfactory. The areas of medium and low QOL based on waste disposal facilities as shown in maps clearly indicates that about 80 per cent of the areas in sample cities are not having regular waste collection and disposal facility.

Problem of Drinking Water

Rajasthan is a State in which shortage of water is a permanent feature. This shortage of water can be seen in urban areas of the State. Most of the cities of Rajasthan are now facing shortage of drinking water, which is a minimum requirement for maintaining the quality of life. Apart from availability of drinking water, an important aspect is the quality of water. One of the tragic aspect of municipal water supply is that sometimes the water supplied is contaminated, causing epidemic. There have been instances of municipal sewerage mixed up with drinking water supply. One of the variable selected for determining the urban QOL is the water supply. Within the city there is a variation in availability of drinking water. This variation is more in cities where available water is lesser. Our study reveals that in a city like Kota, where water is available but its supply is not proper and several areas are not getting sufficient drinking water. This is more true about the slum areas as well as newly developed areas. The same situation is in Bikaner, Sriganganagar, Bharatpur and Tonk. This problem becomes more during summers. Apart from availability, the water supply problem is technical also with the fast growth of urban population and expansion of cities, the technical system has not developed accordingly. Because of this gap some areas of the city are not getting drinking water.

Urban Health

Although, health status of a population is known to differ in accordance with lifestyle and socio-economic conditions. But in urban areas, because of overcrowding population, housing conditions as well as sanitation and other problems, general health problems are common. These problems are more in areas having low quality of life. These areas are disease prone areas. Studies of morbidity patterns in urban areas have shown the preponderances of gastrointestinal and respiratory problems which are closely associated with the prevailing conditions of environment and sanitation. There is a dominance of infectious and parasitic diseases as well as disease of the circulatory system. Viral hepatitis, Cholera and Enteric fever are common in slum areas of the cities. The most common respiratory diseases in the urban areas are Pulmonary Tuberculosis, Influenza and predominantly infantile diseases such as Diphtheria, Chicken Pox and Mumps. Atmospheric pollution is one of the main causes of the increasing incidence of respiratory morbidity in urban areas. Industries and traffic are the two main sources of atmospheric pollution. Malaria and Filariasis are also common diseases in urban areas. The urban environment offers a wide range of conditions suitable for proliferation of vectors and parasites as well as for the transmission of disease causing organism. In brief, it becomes clear that human intervention in the ecosystem represented by an urban area, has been responsible for the complex and multiple health problems.

Planning

The process of urbanisation and rapid increase of population in cities of Rajasthan has resulted in degradation of environment and serious shortfalls in housing, public utilities and community facilities. The urban environment has deteriorated, giving rise to shanty towns and slums, heavy population concentrations, uncontrolled land use and inadequate urban services such as water supply, sanitation, drainage and welfare facilities. These problems are responsible for lower quality of life in urban areas, therefore, planning for the urban development and improvement of QOL is necessary. Apart from infrastructural and spatial planning, social planning approach should be adopted for the improvement of urban QOL. Three general approaches to planning can be delineated:

1. Conventional approaches, which assume that problems can be solved by existing mechanism;
2. Reformist approaches, which assume that the system needs some major modifications, and
3. Radical approaches, which assume that problems cannot be solved by the existing social system.

While suggesting planning strategies for the improvement of environmental quality and QOL, there is a need to combine the first two approaches in the existing social system. The proposed suggestions for planning are based on field study as well as some general suggestions have also been given.

Management for Pollution Control

In urban centres of Rajasthan, especially in cities having above one lakh population, pollution has become the main problem. Various type of pollution, i.e. air, water, land, noise and their spatial pattern have been discussed in selected cities. The study reveals that both immediate and long term steps should be taken to control the pollution. It is also necessary for maintaining the urban QOL.

There is a need to control each type of pollution through proper management. The measures taken to control air, water, land and noise pollution needs separate management system on the one hand, and an integrated system on the other. The following steps may be helpful in integrated management of pollution in cities:

(i) Strengthening of monitoring network to be able to establish periodically the quality status of water bodies and ambient air along with regulating the compliance of emission norms from polluting sources.

(ii) Environmental information system should be established where all types of data will be available.

(iii) There is a need to have sizeable technically and scientifically trained personnel capable of handling problems related to pollution in cities.

(iv) The polluters should be enlightened to made conscious of their social obligation commensurate to profit motives.

(v) Mass awareness programmes should be conducted and full scientific knowledge should be given to the people not only regarding adverse effect of pollution but also for its control.

(vi) Development of 'green belt' around cities is necessary. Although, in master plans of each city, there is a provision for green belt. But the same has not been followed in any city of Rajasthan. Since, cities are expanding at a fast rate, there must be a series of green belt for each new colony.

(vii) Environmental research, including pollution control should be encouraged at all levels. In higher educational institutes, research should be done for local problems of pollution and their results should be taken into consideration by local authorities.

Apart from above measures steps should be taken for the control of each type of pollution. The following measures will be useful for air, water, land and noise pollution:

(a) *Air Pollution*: In big cities of Rajasthan the main cause of air pollution is the industries and automobiles. This should be checked by adopting following measures:

   (i) New industrial sites should be located at a certain distance from residential areas.

   (ii) The selection of industrial site should be done with consideration of air direction.

   (iii) In industries, arrangements for pollution control should be done only after full arrangements of effluent treatment is done, permission for production should be given.

   (iv) Air pollution by automobiles can be reduced by technological innovations, as well as by regular services of the vehicles, replacement of old engines, use of filters, mixing of antitibs in diesel and petrol, etc.

(v) In cities, parking places should be developed along main roads and commercial sites.

(vi) In several cities like Kota, Bikaner, Bharatpur, Tonk, Sriganganagar, etc. national highways are passing and have become a main cause of air pollution. In all cities bypasses should be developed to avoid heavy traffic.

(vii) There must be a 'green belt' around every city. Apart from green belt wherever possible trees should be developed around industries, along roads and also in open spaces.

(b) *Water Pollution*: All the cities of Rajasthan are having water pollution problem, which need immediate attention both by local authorities as well as by individuals. Apart from technological methods there are other measures which are useful in minimising water pollution. Some of them are as follows:

(i) All the domestic and municipal effluents must be drained in water bodies only after treatment.

(ii) The drinking water sources and its nearby area must be kept clean and regular cleaning must be done.

(iii) There must be strict regulations for industrial effluents, that they must not discharge highly toxic water into water bodies. Every industry should develop its own effluent treatment plant.

(iv) The most important aspect would be to educate people not to pollute water. This awareness will help in solving this problem to a great extent.

(v) The treatment of sewage is one of the important measures, which aims in the removal of BOD, phosphorous, nitrogen, solids and bacteria.

(vi) In several localities, there is a problem of water pollution due to absence of proper drainage facilities this should be checked by local bodies and by residents of that area.

(vii) Pollution in local water bodies is also common. For example, in Kota city, river Chambal has become a polluted river because several drains used to pour dirty and waste water into the river water. A plan to control water pollution in river Chambal has already been approved by Central government. Such plans should not only be implemented in time, but steps should be taken in such a way that further pollution may not occur.

(c) *Land Pollution*: In cities land pollution is the result of domestic, municipal and industrial wastes. The measures of waste collection and disposal will be discussed in another heading in this chapter itself.

(d) *Noise Pollution*: Although noise pollution is not a serious problem in cities of Rajasthan. But it is increasing day by day. Therefore, measures for its control should by taken. Some of the measures are as follows:

(i) Noise producing industries should be located away from residential areas.
(ii) Inside industries proper arrangements to minimise noise should be done by constructing soundproof walls.
(iii) Use of horn should be minimum and pressure horn should be banned.
(iv) The noise created by musical instruments, and other indoor equipments can be checked by individuals.

Urban Wastes Management

One of the greater challenges thrown up by rapid urbanisation is the management of urban wastes, both solid and liquid, which are the biggest potential source of pollution and health hazard. The QOL as affected by waste collection and disposal facilities in sample cities has already been discussed in detail. The study reveals that collection and disposal facilities in cities of Rajasthan are not satisfactory and require immediate attention. Although, during last 20 years efforts have been done by local bodies, but because of financial strains as well as due to mismanagement, the problem of waste disposal is growing day by day. It is necessary that emphasis should be given to:

(i) Building of sound organisation under a unitary control;
(ii) Suitable changes in the methods of collection, transport and disposal should be done to meet ever-changing character of city refuse; and
(iii) More hygienic and economical methods of disposal of waste should be adopted on scientific lines.

The maintenance and cleaning of public places is done by the local authorities with powers vested in them under the municipal laws. It is the primary duty of municipalities/municipal councils to make reasonable provisions for the following matters:

(i) Cleaning public places, streets and all spaces, whether such spaces are visited in the Boards or not;
(ii) Removing filth, rubbish, nightsoil, odour, or any other noxious matter from privies latrines, urinals, cesspools, or other common receptacies for such matter in or pertaining to a building or buildings;
(iii) Acquiring, maintaining, changing and regulating places for disposal of dead and of the carcasses of dead animals;
(iv) Constructing slaughter houses, drains, sewers, drainage works, sewerage works, baths, washing places; and
(v) Constructing public latrines, privies and urinals, etc.

In fact, local authorities should act in an effective manner for making all arrangements of waste disposal. Apart from this, it is the duty of every resident of the city to help in waste collection and disposal. The work of waste collection and disposal can also be done on public-private partnership basis.

Developmental Activities in Slum Areas

Existence of slums or *Kachhi Bastis* are common in urban areas of Rajasthan as in case of all Indian towns. The slum areas of the cities in present study have been identified and the nature of QOL in these areas has already been discussed in detail. The intensity of slum is very high in sample cities, i.e. Kota, Bikaner, Sriganganagar, Bharatpur and Tonk. The same is the story of all the towns of Rajasthan having above one lakh population. In these areas quality of life is low. Slums are the marginal areas of cities where the services and facilities are poor, people are illiterate, drinking water supply is miserably on lowest ebb, the housing and street conditions unsatisfactory, etc.

The government both Central and State are very particular about improvement of conditions in these areas and several programmes have been launched. These efforts came to be known as the Slum Improvement Programme (SIP). The programme envisages to provide houses, lavatories, potable water, drains, roads and street lights in slum areas. As early as in 1972 a central scheme of Environmental Improvement of Urban Slums (EIUS) was introduced. But this programme does not provide the long term solution to the problem. In 1996 another scheme known as National Slum Development Programme (NSDP) was launched for providing adequate and satisfactory water supply, sanitation, primary education, health care, etc. to the urban poor. The National Slum Policy 1999 endorses the strategy of upgradation and improvement in all slums. Similarly several other programmes have been started and still continuing but QOL in slums is still below the average. This is because of the fact that in slum areas the growth of population is very high and migration from rural areas creates the worst situation. The following steps may be helpful for the improvement of conditions in slum areas:

(i) Improving employment opportunities for all through an urban employment guarantee scheme.
(ii) Building better quality of housing as a basic human need.
(iii) Improving the quality and quantity of water supply.
(iv) Improving sanitation and sewerage.
(v) Providing better education and health services.
(vi) Ensuring a minimum calorie and protein intake for all families, particularly all children and the girl child.

Urban Health

The quality of life in cities can be improved by adopting health care measures. The following elements make up the strategic framework for health promotion in cities:

(a) Promoting health public policy,
(b) Creating supportive environments,
(c) Strengthening community participation,
(d) Improving personal skills, and
(e) Reorienting health care delivery system.

The urban health programme should focus on promoting and sustaining local health action based on the primary health care approach. The urban health care strategy should be based on basic principal *'Health for All'*. This can be jointly done by government as well as people. The need of the time is that the government, through appropriate policies and programmes, should facilitate this capacity building process at the earliest possible.

Integrated and Spatial Approach for the Planning of QOL

As discussed earlier QOL is based on several variables, i.e. social, economic, cultural, etc. each variable has its own importance and due consideration should be given to each one while planning the QOL. But more important is that planning should be done in an integrated manner with consideration of spatial aspect. Because area as a whole or its part should be developed and when an area is developed, the people will develop and naturally quality of life will develop. Prior to spatial planning a brief reference of Huxley's model of culture and quality of life is useful.

Julian Huxley, a British Biologist, has developed a model which shows relationship between culture, biological evolution of man and QOL. This becomes clear from Figure 8.1.

The three components suggested by Huxley are menti-facts, socio-facts and artifacts.

Figure 8.1 *Huxley's Model of Culture and Quality of Life*

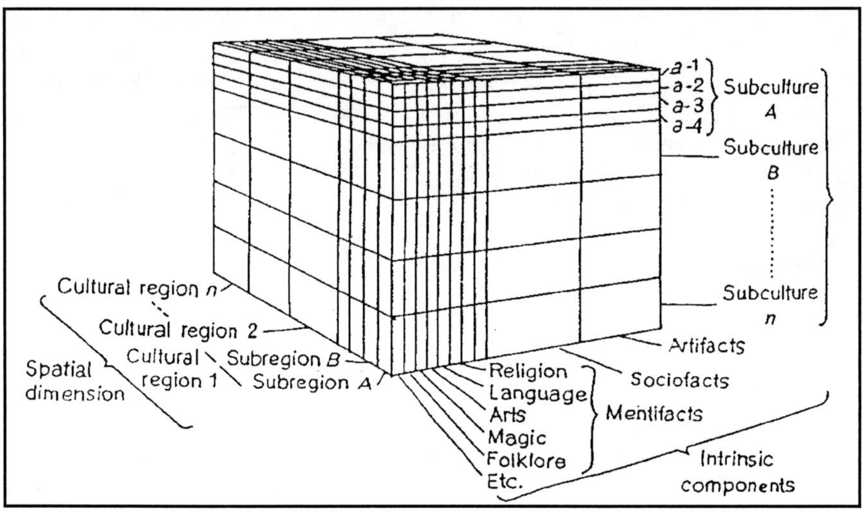

Menti-Facts: These are considered in terms of religion, language, magic, folklore, artistic tradition and others. These are crucial and durable elements of culture, because they relate to human ability to think, to forge ideas and images against which other aspects of culture could be measured.

Socio-Facts: These aspects of culture create link between individual and groups as well. At individual level they include family structure, sexual behaviour and child rearing. At group level they include political and educational system.

Artifacts: These are known as cultural freight of material manifestation of culture. These include group of material technology, e.g. food, shelter, transport, besides land utilisation, crop production and design on the clothes are known as artifacts.

The above model shows a generalised pattern of the cultural growth and its related issues including QOL.

The best approach of planning for the improvement of QOL is the integrated spatial approach. In present study urban QOL has been determined on the basis of housing conditions, street conditions, facilities of drinking water, waste disposal and drainage. Both QOL based on each variable and based on composite values have been determined and their spatial pattern has been discussed. The determination of QOL on the basis of composite values is a part of integrated approach because all the factors have been taken into consideration. Another significant aspect is that areas of high, medium and low QOL have been identified in sample towns. These areas will form the basis of planning the QOL in urban areas. Accordingly, priority should be given to the areas of low quality of life in each city. These areas are where minimum facility of housing, drinking water, waste disposal, drainage are not available and street conditions are worst. Therefore, for each such area development plan should be prepared. For the areas of medium quality of life problems of each area should be identified, as done in case of sample cities and for solving these problems steps should be taken. As regard areas having high quality of life, regular works should continue, because these areas are developed areas and they will continue to develop with routine developmental works. The need of the time is that the gulf between areas of high, medium and low QOL should be minimised, so that everyone is able to live in a healthy environment.

References

Bose, Ashish (1955), 'Urbanisation and Slums', in Prodipto Roy and Shangon Das Gupta (eds), *Urbanisation and Slums*, Har-Anand Publications, New Delhi, p. 19.

Francis, Cherunilam (1984), *Urbanisation in Developing Countries*, Himalaya Publishing House, Bombay, pp. 174–76.

Wishwakarma, R.K. (1984), *Urban and Regional Planning Policy in India*, Uppal Publishing House, New Delhi, pp. 174–76.

9 Summary and Conclusions

The present study entitled, '*Urbanisation, Environmental Degradation and Quality of Life*' is related with three aspects, i.e. (i) Urbanisation, (ii) Environmental Degradation, and (iii) Quality of Life. All the three aspects are although, independent but interrelated with each other and have had direct impact on urban development. Urbanisation is a process which relates to concentration of people in non-agricultural occupations and concentration of non-agricultural land use in specialised area and it is also an important aspect of the process of economic and social development. At the same time, urbanisation is a main cause of degradation of environment. Urbanisation has bought with it many problems, including shortage of housing, inadequate water supply, sanitation, waste disposal facilities, congestion, traffic problems, pollution as well as unsafe social environment. All these urban problems are responsible for lower quality of life in many parts of the cities. All this supports the interrelationship of the focal theme of the present study that urbanisation is a cause of environmental degradation and environmental degradation is responsible for lower quality of life in many areas of the cities.

Summary

The present study is an attempt to investigate the dynamic relationship between urbanisation, environmental degradation and quality of life in cities of the Rajasthan State. The important aspects discussed in present study are:

(a) Geographical individuality of the region;
(b) Trends of urbanisation;
(c) Degradation of environment in Rajasthan;
(d) Spatial pattern of pollution in urban areas;
(e) Spatial pattern of quality of life in urban centres;
(f) Levels of quality of life; and
(g) Problems and planning.

The region selected for present study is the Rajasthan State, which is the largest State of the Indian Union with 3,42,239 sq. km area (i.e. nearly 10.43 per cent of India). The Rajasthan State is a unique historico-geographical region and having diversified economy primarily based on agriculture, mining, livestock and to some extent on industry. According to 2001 census 56.47 million of Rajasthan constituted about

5.50 per cent of India's total population. The percentage of urban population is 23.88, residing in 222 towns.

The subject selected for present study is interdisciplinary in nature. Although, our study is based on geographical analysis but it is also related with subjects like sociology, urban planning, economics and environmental studies. The subject is important both from theoretical as well as applied point of view and will also be helpful to administrators and policy makers for the improvement of urban QOL.

In Rajasthan urbanisation was started during ancient period and this process continues in different historical periods. But it was only in the post-independence period that the phase of industrialisation, urbanisation and socio-economic transformation got the real start. Expansion in administrative needs, industrial and commercial activities and development in communication gave new impetus to the growth of new towns. Spectacular increase in urban population took place between 1941 and 1951 with a net growth rate of 17.46 per cent from 5.22 per cent between the decades of 1951 to 1961 showing again a tremendous increase of net 20.22 per cent increase in 1981. There was a corresponding increase in number of towns as well. From 1951 to 1991 the state's urban population multiplied by seven times growing from 29.55 lakh to 100.67 lakh. The percentage of urban population to total population in 2001 was 23.88. The number of towns has increased from 135 in 1901 to 222 in 2001. The number of towns having more than one lakh population is 20. The economic generators promoting urban growth and process of urbanisation in the state are agriculture, minerals, industries, tourism, trade and commerce. The process of urbanisation is determined by (i) environmental conditions, (ii) agricultural development, (iii) mining development (iv) industrial development, (v) development of amenities, (vi) infrastructural development, (vii) institutional development, and (ix) social development. The urbanisation in Rajasthan has given new boost to the economic development, but at the same time it is also a cause of degradation of environment as well as lower QOL.

Degradation of environment has emerged as a major concern for the State of Rajasthan. This is more a problem for urban areas. The problem of environmental degradation in Rajasthan has been discussed under two heads (i) general environmental degradation, and (ii) urban environmental degradation.

The major problems of environmental degradation in Rajasthan are:

(i) Desertification and land degradation
(ii) Drought
(iii) Deforestation
(iv) Scarcity of water
(v) The fluoride problem
(vi) Salinity and water logging, and
(vii) Loss of biodiversity and wildlife.

On the other hand main problems of urban environmental degradation are:

(i) Lack of sanitation facilities
(ii) Problem of urban waste disposal
(iii) Industrial pollution
(iv) Air pollution by automobiles, and
(v) Water pollution, etc.

All these problems have been discussed with reference to cities of Rajasthan.

Spatial analysis is the prime concern of geographical study, therefore, spatial pattern of pollution in urban area has been discussed in detail. The spatial pattern of air, water, land and noise pollution in sample cities has been examined. The samples selected for the study are Kota, Bharatpur, Sriganganagar, Tonk and Gangapur city. Areas of high, medium and low pollution have been identified for air, water, land and noise pollution and the same has been depicted on the city map of selected towns.

The QOL in urban centres of Rajasthan has been analysed in detail. The QOL is a relative term, varies from nation to nation, society to society as well as from person to person, therefore, it is very difficult to develop a uniform criterion for its analysis. As discussed in Chapter–1 and 6, several scholars including geographers and a few agencies have developed and applied certain variables for the measurement of QOL. But all these studies have been done with reference to developed countries, where conditions are entirely different from countries like India. For present study a framework for the determination and measurement of QOL has been designed. Accordingly, the spatial pattern of urban QOL has been determined on the basis of following variables:

(i) Housing conditions
(ii) Nature of roads
(iii) Drinking water facilities
(iv) Internal drainage system, and
(v) Waste disposal facilities.

A detailed fieldwork has been conducted in sample cities and identification of areas having high, medium and low QOL has been determined by assigning weights for each category. The five cities selected for this study are Kota, Bikaner, Sriganganagar, Bharatpur and Tonk. For each city wardwise status of QOL based on each variable has been determined and maps to show QOL based on selected variables have been prepared for detailed analysis.

In the next stage, levels of QOL or hierarchy of the urban QOL has been examined on the basis of methodology described in Chapter–7. The QOL for all wards of the city has been determined on the basis of composite index value. The index value for each variable has been given and finally on the basis of total index values or composite index value, status of ward in terms of QOL has been determined and the same has been depicted on city maps. This overall analysis provides a basis for levels of urban QOL, i.e. high, medium and low and this identification finally helps in city planning and for the planning of the improvement of QOL.

Finally, problems of urbanisation, environmental degradation and QOL have been discussed and suggestions for planning have been given. The problems related to housing, urban slums, sanitation, drainage and sewerage system, waste disposal, drinking water, urban health have been discussed with reference to sample towns of the study area. A framework for planning has also been suggested for the management of pollution, waste management, development activities in slum areas, urban health, etc. an integrated and spatial approach for the planning of urban QOL has also been given. In brief, the study unfolds the entire urban structure in terms of environmental degradation and quality of life.

Conclusions

In entire study, conclusions have been drawn in each chapter for concerned aspect, i.e. urbanisation, degradation of environment and quality of urban life. The main conclusions of the study are as follows:

1. Rajasthan has a long standing history of urbanisation and planned settlements dating back to 2500 BC.
2. The actual phase of urbanisation has started only after independence.
3. The urban centres of the State had strong influence of social, economic and political environment prevailing in the State during medieval, colonial and post-independence periods.
4. In spite of the challenges posed by extreme climatic conditions and geographical diversities in this large State, efforts of planned development of cities have been taken, which have shown encouraging signs in terms of infrastructure development and supplementing social sector goals.
5. Major economic generators promoting urban growth in the State are agriculture supplemented by improved irrigation in canal command area, mining and mineral-based industries, industrial development, transportation, trade and commerce and tourism industry.
6. More than half of the population of the State is concentrated in 14 class I towns of the State.
7. Quality of environment in urban areas has deteriorated over the period of time.
8. Pollution is the main problem of environmental degradation in cities of Rajasthan and pose serious threat to the life of inhabitants.
9. Among the most important causes of increasing pollution are fossil fuel burning, increasing vehicular traffic, and disposal of solid and liquid waste without treatment.

10. Existing system of generation, collection and disposal of the solid waste is grossly unsatisfactory and is ill-equipped to meet the changing consumption habits that accelerate the face of solid waste generation.
11. In Rajasthan, desertification, land degradation, drought, deforestation, scarcity of water are major problems of environmental degradation, which should be taken as a priority by the government and planned efforts should be done in solving these problems.
12. The cities of Rajasthan are facing problems of air, water, land and noise pollution as well as problems of sanitation, waste disposal, drinking water, etc. These problems vary from city to city, therefore, for each city a separate plan should be prepared to control these problems. Specific problems have been highlighted in case of Kota, Bikaner, Sriganganagar, Bharatpur and Tonk.
13. Another aspect of urban pollution is their spatial pattern, i.e. within the city there are certain areas which are highly polluted, while others are lesser in terms of air, water and land pollution. Such areas have been identified for sample cities. Priority should be given to improve conditions of those areas which are highly polluted.
14. In Indian conditions QOL can be determined on the basis of following variables: housing conditions, street conditions, drinking water facilities, internal drainage facilities and waste disposal facilities.
15. The spatial pattern of urban QOL on the basis of each above mentioned variable is to be determined.
16. As done in Chapter–6 of the present study areas of high, medium and low QOL in terms of housing, street conditions drinking water, internal drainage and waste disposal facility have been determined for sample cities.
17. Such identification of areas in city wards helps city planners to solve problems associated with specific areas.
18. Apart from specific problems, levels of QOL in urban centres are to be determined and their spatial pattern for each city should be analysed, as done in Chapter–7 of the present study.
19. The levels of QOL in cities are based on composite index value obtained through total index value of all the variables for each ward/area.
20. The wardwise depiction of QOL on maps has been done and also their characteristics have been discussed, which indicates the problems associated with specific areas of the city.
21. Slums, now known as '*Kachhi Basti*' have developed in all the cities of Rajasthan. It has been observed that bigger is the city, more are the slums.
22. Slums are the overcrowded and unauthorised settlements on public/private lands, situated along railway tracks, open drains, river banks, under bridges and the flyovers and/or open spaces without any tenural rights.

23. These slums reflect the darker and lesser known aspects of the unplanned urbanisation process.
24. All the slums are having lower QOL and highly disease prone areas.
25. Several efforts have been done for the improvement of the conditions by Central and State governments but because of fast growth of population in slum areas and also due to migration of rural people in cities, the conditions of slum dwellers is still not satisfactory.
26. There is a need to develop leadership training programmes for slum dwellers. There is an urgent need to literate slum dwellers and also to develop awareness among them regarding health, hygiene, family planning, etc.
27. The study reveals that urban centres of Rajasthan are facing problems of housing, slums, sanitation, sewerage, waste collection and disposal, drinking water, health as well as air, water, land and noise pollution.
28. The above mentioned problems are responsible for lower urban quality of life.
29. There is a need to adopt an integrated and spatial approach for the improvement of urban QOL. An action plan should be prepared with a challenging vision and specific actions and a concrete target to improve the QOL.

The choice for the future is not between planning and no planning, but rather on what level it should be done. Whatever our formal plans for the city of the future, much of what will actually happen is the result of untold numbers of diverse decisions made by different individuals. As stated by Jane Jacobs (1961: 241):

> 'Most city diversity is the creation of incredible numbers of different people and different private organizations with vastly differing ideas and purposes, planning and contribution outside the formal framework of public action. The main responsibility of city planning and design should be to develop – in so far as public policy and action can do so – cities that are congenial places for this great range of unofficial plans, ideas and opportunities to flourish, along with the flourishing of the public enterprises'.

The condition of environmental degradation in cities of Rajasthan is not critical and it can be improved by joint action of the government and public. Planning is necessary so as the peoples' participation to control degradation of environment and improvement of QOL, because cities will continue to grow and grow at a rapid rate.

Reference

Jacobs, Jane (1961), *The Death and Life of Great American Cities*, Vintage/Random House, New York.

Bibliography

Ali, Sabir (ed.) (2006), *Dimensions of Urban Poverty*, Rawat Publications, Jaipur.

Bederman, S.H. (1974), 'The Stratification of "Quality of Life" in Black Community of Atlanta, Georgia', *Southeastern Geographer*, 14, pp. 26–37.

Bell, David and M. Jayne (2006), *Small Cities*, Routledge, New York.

Berry, B.J.L. (1973), *The Human Consequences of Urbanisation*, Macmillan, London.

Bhatnagar, K.K. and K.K. Gadeock (eds) (2007), *Urban Development and Administration*, Aalekh Publishers, Jaipur.

Bhatt, K.N. (ed.) (2008), *Population, Environment and Health*, Rawat Publications, Jaipur.

Bhattacharya, B. (2006), *Urban Development in India*, Concept Publishing, New Delhi.

Boone, C.G. and A. Modarres (2006), *City and Environment*, Temple University Press, Philadelphia.

Cox, K.R. (1972), *Man, Location and Behaviour: An Introduction to Human Geography*, John Wiley & Sons, New York.

Dickinson, G. and K. Murphy (2007), *Ecosystems*, Routledge, New York.

Dickinson, J.C., R.J. Gray and D.M. Smith (1972), 'The "Quality of Life" in Gainesville, Florida: An Application of Territorial Social Indicators', *Southeastern Geographer*, 12, pp. 121–32.

Drewnowski, J. (1970), *Studies in the Measurement of Levels of Living and Welfare*, UN Research Institute for Social Development, Geneva.

Dutta, Amlan and M.M. Agrawal (eds) (1992), *The Quality of Life*, B.R. Publishing Corporation, Delhi.

Eyles, J. (1974), 'Social Theory and Social Geography', *Progress in Geography*, 6, pp. 27–87.

Fakhruddin (1991), *Quality of Urban Life*, Rawat Publications, Jaipur.

Gatwell, A.C. (2005), *Geographies of Health*, Blackwell, Malden, USA.

Ghosh, A. (2003), *Urban Environmental Management*, Concept Publishing, New Delhi.

Goudie, A. (2006), *The Human Impact on the National Environment*, 6th edition, Blackwell, Malden, USA.

Graves, P.E. (2010), *Environmental Economics*, Rawat Publications, Jaipur (Indian Reprint).

Hesselberg, J. (2002), *Issues in Urbanisation*, Rawat Publications, Jaipur.

Howe, G.M. (1972), *Man, Environment and Disease in Britain*, David and Charles, Newton Abbot and Barnes and Nobel, New York.

Jana, M.M. (2003), 'Basic Factors for Deteriorating the Quality of Life in India', *Geographical Review of India*, 65 (4), pp. 328–43.

Kayastha, S.L. (1993), 'Environment, Development and Quality of Life', *Annals of the NAGI*, XIII (2), pp. 54–75.

Knox, P.L. (1975), *Social Well-Being: A Spatial Perspective*, University Press, Oxford.

Koelle, H.H. (1974), 'An Experimental Study on the Determination of a Definition for the "Quality of Life"', *Regional Studies*, 8, pp. 1–10.

Kumar, P. and B.S. Reddy (2007), *Ecology and Human Well-Being*, Sage Publications, New Delhi.

Liu, B.C. (1973), *The Quality of Life in the United States 1970: Index, Rating and Statistics*, Midwest Research Institute, Kansas City, MO.

Mandal, R.B. (2000), *Urban Geography*, Concept Publishing, New Delhi.

McGillivray, M. and M. Clarke (eds) (2006), *Understanding Human Well-Being*, Bookwell, New Delhi.

Mohanty, L.N.P. and S. Mohanty (2005), *Slum in India*, A.P.H. Publishing Corporation, New Delhi.

Mukherjee, Ramkrishna (1989), *The Quality of Life*, Sage Publications, New Delhi.

Mukherji, S. (2004), *Fragile Environment*, Manak Publishing, New Delhi.

Munsi, S.K. (1990), 'Welfare Geography, Poverty and Environment', *Geographical Review of India*, 52 (3), pp. 1–3.

Naik, N.T.K. and S.M. Rahman (2007), *Urbanisation of India*, Serials Publishing, New Delhi.

Nath, S.K. (1973), *A Perspective of Welfare Economics*, Macmillan, London.

Nigar, Anjum (1997), 'Habitat, Environmental Degradation and Quality of Life in Modi Nagar', *Geographical Review of India*, 59 (2), pp. 159–63.

Ontell, R. (1973), *The Quality of Life in San Diego*, The Urban Observatory of San Diego, San Diego.

Palen, J.J. (2008), *The Urban World*, Paradigm Publishers, London.

Phadke, V.S. and S. Banerjee-Guha (eds) (2007), *Urbanisation, Development and Environment*, Rawat Publications, Jaipur.

Phillips, David (2006), *Quality of Life*, Routledge, London.

Riddle, R. (2003), *Sustainable Urban Planning*, Blackwell Publishing, Oxford, UK.

Roberts, N. (1994), *The Changing Global Environment*, Blackwell, Malden, USA.

Saxena, H.M. (2004), *Environmental Geography*, Rawat Publications, Jaipur.

Sinha, Rekha and U.P. Sinha (2007), *Ecology and Quality of Life in Urban Slums*, Concept Publishing, New Delhi.

Small, J. and M. Witherick (1995), *A Modern Dictionary of Geography*, Edward Arnold, London.

Smith, D.M. (1973), *The Geography of Social Well-Being in the United States*, McGraw-Hill, New York.

Smith, D.M. (1977), *Human Geography: A Welfare Approach*, Edward Arnold, London.

Sutton, P.W. (2007), *The Environment Polity*, Polity Press, Cambridge.

Terleckyj, N.E. (1975), *Improvements in the Quality of Life: Estimates of Possibilities in United States*, National Planning Association, Washington, D.C.

Thakur, B. and Others (2005), *Urban and Regional Development in India*, Vol. I and II, Concept Publishing, New Delhi.

Whitehead, M. (2007), *Spaces of Sustainability*, Routledge, New York.

WHO QOL Group (1995), 'The World Health Organisation Quality of Life Assessment: Position Paper from the WHO', *Social Science and Medicine*, 41, pp. 1403–09.

Wright, R.T. (2007), *Environmental Science*, 9th edition, Prentice Hall, Delhi.

Index